People, Cultures and Societies: Exploring and Documenting Diversities

The series proposes to capture the diversities in people and their communities in India. It provides a unique and innovative resource for anthropological knowledge, philosophy, methods, and tools to understand, analyse and formulate sustainable, innovative solutions to address socio-cultural issues in India.

India is a repository of varied cultures and diversities. With the globalisation and development process, the cultural fabric is changing. Customs, traditions, beliefs, on one hand, food habits, art and craft, weaving, dyeing, and handloom artefacts, on the other, are undergoing a metamorphosis. It is imperative to explore, understand and document the process of changing diversity and relational inequalities. The series encompasses richness in art, craft, language, dance, music, folklore, food culture and beliefs, traditions and practices. It addresses the issues of development disparities, inequality, and segregation on the axes of caste, class, religion, region, ethnicity, and gender. The series publishes methodologically rigorous and theoretically sound, critical and comparative, empirical research peer-reviewed volumes related to non-codified healing practices, gender-based violence, migration induced vulnerabilities, child abuse, social identity-based work on a national, regional and local level, welcoming case studies, as well as comparative and applied research.

The series is of interest to the academicians and students in the discipline of sociology, anthropology, psychology, social work, history, philosophy, and public health, and to all of those interested in a wide-ranging overview of art, culture, and politics. It accepts monographs, edited volumes, and textbooks.

K. M. Ziyauddin

Occupational Caste Groups' Illness and Health in India

Lessons from the Lives of Conservancy Workers

ANTHROPOS INDIA FOUNDATION

Springer

K. M. Ziyauddin
Department of Sociology
Maulana Azad National Urdu University
Hyderabad, Telangana, India

ISSN 2662-6616 ISSN 2662-6624 (electronic)
People, Cultures and Societies: Exploring and Documenting Diversities
ISBN 978-981-96-8966-8 ISBN 978-981-96-8967-5 (eBook)
https://doi.org/10.1007/978-981-96-8967-5

This Springer imprint is published by the registered company Springer Nature Singapore Pte Ltd.
The registered company address is: 152 Beach Road, #21-01/04 Gateway East, Singapore 189721, Singapore

I dedicate this book to the Hadi Community for their rich history of living and showing an experiential lifeworld.

&

My children, Adifaah and Aaraf, and my spouse, Dr. Hena Kauser

I hope that this reaches all who have not studied about the caste Hadis (a Dalit, unread and unwritten) of India.

Foreword

In the rich tapestry of India's diverse social fabric, the threads of occupational caste groups often reveal stories of resilience, struggle, and unyielding spirit. 'Occupational Caste Groups' Illness and Health in India: Lessons from the Lives of Conservancy Workers' offers a unique and poignant exploration of one of the nation's most marginalised communities. This book brings to light the health and illness experiences of conservancy workers, a perspective often overlooked yet crucial for understanding urban sanitation and public health in India. Despite their challenging circumstances, the resilience and unyielding spirit of these workers are truly inspiring and filled with socio-cultural discrimination.

Conservancy workers, responsible for maintaining cleanliness and hygiene in our cities, perform physically demanding and socially stigmatised duties. These individuals, frequently belonging to lower caste groups, endure significant health risks due to their occupation. Despite these challenges, their role is not just essential but also invaluable to urban sanitation and public health. Their contribution, often underappreciated, cannot be overstated, and their health concerns mustn't be neglected. Understanding and appreciating their role is crucial for improving their working conditions and overall well-being.

The author of this book has undertaken a critical and empathetic examination of the lives of conservancy workers, offering an in-depth analysis of the intersection between caste, occupation, and health. Through meticulous research and compelling narrative, this work sheds light on the systemic discrimination and occupational hazards these workers face and underscores the urgent and pressing need for immediate policy reforms and health interventions. The urgency of this need should concern us all, and it should motivate us to advocate for change. These reforms can improve their working conditions and overall well-being. Health perceptions are essential to study, along with sociological evidence and data illustrating how experiences are pertinent to the meaning of illness and health. The qualitative data emerging from the intensive fieldwork in this book analyses everyday social interactions of the conservancy workers with the health care personnel, health administration, physicians, and health care providers stand out to be significant aspects and the least touched dimension in the sociology of health and illnesses studies in India.

Readers could utilise this book regarding caste and occupation concerning health perceptions and health care accessibility by the community. The myriad forms of household members' engagements within themselves and how children, women, and the elderly decide to access health care are engaging in this study.

Prof. Vivek Kumar
Professor of Sociology & Prof.
Ambedkar Chair in Sociology
Centre for the Study of Social Systems,
School of Social Sciences
Jawaharlal Nehru University
New Delhi, India

Preface

The Hadi caste is a historically disadvantaged minority within Hindu society. It is predominantly located in the eastern region and states of India, specifically in the states of West Bengal, Odisha, and Jharkhand. Traditionally, jobs of lower social standing have been associated with the Hadis. India frequently categorises them as Scheduled Castes (SC), a term that recognises and provides preferential treatment to communities that have historically faced disadvantages. Historically, the Hadis have performed cleaning, sweeping, and scavenging duties. These professions have positioned them at the lowest level of the social hierarchy, exposing them to considerable social shame and discrimination. The inflexible caste system in India has kept them marginalised, limiting their ability to get an education, economic prospects, and social advancement. From an economic standpoint, the Hadis have frequently experienced destitution, with restricted opportunities for property ownership, access to resources, and secure work. Their conventional livelihoods fail to provide economic stability, and the absence of educational prospects has deepened their financial difficulties. Many individuals in the Hadi community still engage in low-skilled occupations, although some have endeavoured to find alternate economic opportunities in urban regions. The Hadi caste has had significant difficulties accessing education. Financial limitations and societal prejudice have traditionally restricted their ability to attend educational institutions. Nevertheless, government measures targeted at enhancing the education of Scheduled Castes have started to have an impact in recent years. Through the implementation of scholarships, reserved seats in educational institutions, and other affirmative action programmes, members of the Hadi community have been able to access higher education opportunities and enhance their socio-economic standing.

The Hadi population has frequently experienced substandard living conditions with insufficient housing, sanitation, and health care accessibility. Engaging in their customary professions exposes people to several health hazards, such as respiratory issues, infections, and other occupational ailments. Government initiatives targeting improving living standards for Scheduled Castes have yielded mixed results, as many Hadis continue to confront substantial health and sanitation obstacles. The Hadi group possesses distinct cultural customs, ceremonies, and societal behaviours. Notwithstanding their difficulties, they retain a diverse cultural legacy, encompassing traditional music, dances, and celebrations. Their cultural identity plays a crucial role in their community, fostering a feeling of oneness and ongoing existence.

The Hadi caste in modern India faces significant social and economic difficulties. Although affirmative action programmes have yielded certain advantages, the inherent and entrenched character of caste-based discrimination hampers the pace of advancement. Advocacy groups and NGOs actively raise public awareness and promote the rights and well-being of the Hadi community. The Hadi caste exemplifies the long-lasting influence of India's caste system. Although facing considerable challenges, the community actively pursues improved possibilities and increased social recognition. It is crucial to continue tackling the economic, educational, and health issues faced by the Hadis and other marginalised groups to ensure they attain a fair and impartial society.

Hyderabad, India K. M. Ziyauddin

Acknowledgements

The wealth of data and scholarly articles that underpin this dissertation would not have been possible without the insightful guidance of Prof. Tulsi Patel. Her unwavering commitment to enhancing my writing skills and her profound advice were exceptional, and I am deeply thankful for this. I also acknowledge the steadfast support and academic interactions. I would also like to pay my respects to Prof. Abhijit Dasgupta, Prof. Roma Chatterji, Dr. Shashi Bhusan Singh, Prof. Sanghmitra Acharya, Prof. Abdul Matin, and Prof. Arvinder Ansari, Prof. Kancha Ilaiah, Prof. Biswajit Das, Prof. Mohammad Akram, Prof. Vivek Kumar, Prof. Azra Abidi, Dr. Gomati Bodra, Prof. G. Nagaraju. My teachers, Prof. Mohammad Talib (at Oxford) and Prof. Biswajit Das at JMI, and the other two whom I lost during COVID-19, Prof. Mohini Anjum and Prof. Ravi Kapoor, for their constant moral support, and they wanted me to publish this work sooner whenever I had a telephonic conversation with them. Professor Vivek Kumar, whom I am obliged for his sociological insights in pointing out the social facts as phenomenon found in the lives of Hadis in the process of seeking treatment for illnesses and questions to the accessibility of healthcare and so to other matters that came only by unstinted readiness to read my manuscript to write a meticulous forward. I am grateful and much gratitude to him for this academic support.

I sincerely thank the Department of Sociology at Delhi School of Economics, Ratan Tata Library, JNU Library, Jamia Millia Islamia, University of Hyderabad and MANUU Library. These institutions provided the necessary resources and support for the successful completion of this manuscript whenever needed. I also acknowledge the environment built in the university by the Vice Chancellor, Prof. Syed Ainul Hasan, and officials, including Prof. SK Ishtiaque Ahmed, which helped me complete a vital manuscript, without which this outcome would have been difficult. First and last, the entire Hadi Cooli, including Gandhi Hadi, Santu, Mantu, Dr. Mandal, Dr. Singh, Ajaz Bhai, Rajesh, and my brother Dr. Khwaja Mohammed Tabrez, without whom field re-visit and revised data would not have been possible.

My family's and relatives' love and affection reflect their preference for the familial tie, which is still significant and influential in Indian society, as Prof. Patel sometimes says.

Hyderabad, India K. M. Ziyauddin
2024

Contents

About the Author

K. M. Ziyauddin is an Associate Professor and Head in the Department of Sociology at Maulana Azad National Urdu (central university) University, Hyderabad. His areas of research include exclusionary studies of minorities, the sociology of health and illness, public health and sanitation workers, SBM and Dalit studies, urban space and marginalisation, gender politics, sociology of minorities, and questions of identity and citizenship.

A few of his book publications include *Muslims of India: Exclusionary Process and Inclusionary Measures, Dimensions of Social Exclusion: An Ethnographic Exploration, Minorities in India: Forms and Perspectives, and Sociology of Health in a Dalit Community: Axes of Exclusion*, as well as research papers published in national and international journals. He is an alumnus of the International Visitor Leadership Program of the US Department of State and co-founder of The Democracy Dialogue, India.

Abbreviations

ADM	Administrative Division
ANM	Auxiliary Nurse Midwife
AYUSH	Ayurveda, Yoga, Unani, Siddha and Homoeopathy
BCCL	Bharat Coking Coal Limited
BDO	Block Development Officer
BGH	Bokaro General Hospital
BS City	Bokaro Steel City
BSL	Bokaro Steel Plant Limited
CHC	Community Health Centre
CIL	Coal India Limited
CMC	Chas Municipal Corporation
CMO	Chief Medical Officer
CS	Civil Surgeon
DC	Deputy Commissioner
GOI	Government of India
IAY	Indira Awas Yojana
ICDS	Integrated Child Development Service
ICU	Intensive Care Unit
IPD	In Patient Department
JHYSM	Jharkhand Hadi Yuva Samaj Manch
JMM	Jharkhand Mukti Morcha
KMMHRC	K M Memorial Hospital and Research Centre
MADA	Mines Area Development Authority
MBBS	Bachelor of Medicine and Bachelor of Surgery
MHRC	Muskan Hospital and Research Centre
MLA	Member of Legislative Assembly
MoHFW	Ministry of Health and Family Welfare
MoSJE	Ministry of Social Justice and Empowerment
MP	Member of Parliament
MR	Medical Representative
NCAER	National Council for Applied Economic Research

NFHS	National Family Health Survey
NGO	Non Government Organization
NICU	Neo-natal Intensive Care Unit
NNHRC	Neelam Nursing Hospital and Research Centre
NREGA	National Rural Employment Guarantee Act
NRHM	National Rural Health Mission
OOP	Out of Pocket
OPD	Out Patient Department
OTCM	Over the Counter Medicine
PHC	Primary Health Care
RMP	Registered Medical Practitioner
SAIL	Steel Authority of India Limited
SC	Sub-Centre
SC	Scheduled Caste
SDH	Sub Divisional Hospital
SDO	Sub Divisional Officer
SES	Socio-Economic Status
SKA	Safai Karamchari Andolan
ST	Scheduled Tribe
TB	Tuberculosis
TBA	Traditional Birth Attendant
TH	Traditional Healer
UNHM	Urban National Health Mission

Chapter 1
Introduction

This chapter delves into the perceptions of illness and health, particularly in the context of the caste-based occupation of the Hadis. The Hadis, a Scheduled Caste (SCs) or Dalits, are traditionally engaged in manual scavenging. The research argues that perceptions of illness and health are significantly influenced by factors such as household preferences, the nature and severity of illness, and the potential treatment costs. These factors, in turn, shape the choices made in health care centres. While the primary focus is on understanding the Hadis' perception of illness and health, the thesis also provides an ethnographic description of their jobs and how the community's nomenclature has evolved to 'sweeper'.

The Hadis have resided in the densely populated Hadi Cooli for several generations. I've known Hadi Cooli since I was in school in Chas, Bokaro district, Jharkhand, in the 1990s.

The neighbourhood was characteristically different from other Cooli[1] and *Mohalla*. Filthy streets, roads clogged with drain water, wastewater from domestic use, water from domestic use, etc., such sights left a huge impression on me, and I often reflected on their living conditions. Why was their Cooli more congested than others? I had greater clarity on the structural mechanisms that determine their living conditions when I read a few works like Khare (1984), Massey (1997), and Ilaiah (1998). However, the specificities of their life remained unanswered. They precipitated my research question on illness perception and health-seeking practices among Dalits, for which I already had a site in mind: the Hadi *Cooli* in Chas.

Further, even though the book's title mentions the larger category 'Dalits', this work focuses on a single caste in the Bokaro district. I am aware that Dalits, as a category needs, have been overused indiscriminately and undifferentiated recently and need to be used with precaution (Shah 2002). I have also followed such precautions to avoid generalising the concept of Dalits into a caste, as it would be tantamount

[1] Cooli refers only to Scheduled Castes populated localities in Chas (Bokaro) and *Mohalla* is referred to Muslim localities.

K. M. Ziyauddin, *Occupational Caste Groups' Illness and Health in India*, People, Cultures and Societies: Exploring and Documenting Diversities,
https://doi.org/10.1007/978-981-96-8967-5_1

to undermining the specificities of the Hadis living in Chas town, Bokaro. Moreover, the Hadis have four sub-castes similar to the Bhangis in Rajasthan having four sub-castes namely Bhangi, Dholio-Ke-Bhangi, Hella Bhangi, and Mazbi (Shyamlal 1984). I call Hadis as Dalit in the title, but for the ethnographic research method, the concept of Dalit is not operationalised in the writings.

Hadis, a community of sweepers in the Municipal town of Chas, Bokaro, were notified as one of the Scheduled Castes (SCs) in the Census on India Document, 2001. The Hadi caste is one of the 22 Scheduled Castes residing in Jharkhand. As per preliminary research and findings, first, there has not been much research on this community in the state and their perceptions of illness and health. Second, this caste's traditional engagement in manual scavenging occupation distinguishes them from the other SCs in the state.

Castes like the Hadis are traditionally attached to 'unclean' occupations such as manual scavenging, carrying nightsoil (now changed), and cleaning latrines. These are commonly found occupations that are based on caste hierarchy. A few studies have been carried out to highlight the conditions of Bhangis in Rajasthan (Shyamlal 1984; Chaplin 1997, and Mehtar in Bihar (Sachchidananda 2002). Malavika (1976) and Grover (2007) studied the urban slums of Delhi to understand the status of caste groups engaged in menial occupation. These castes also have poor financial conditions that affect health care capacities.

In common understanding, perception is seeing, experiencing, and consequently having a sense of objects, problems, physical movements, and individuals' experiences (Funch 1995). By experiencing, people notice and understand those senses in their minds and thoughts, which give meaning and translate at their cognitive levels. There is also a consistency in translating the perceived meanings and interpreting them in their languages (Benyamini 2011; Gregory 1987).

Hadis, to an extent, live in a close community borderland and are knitted deeply in their kinship network that does not provide space for others. Their social interaction is limited to their caste members, even though a few elders visit Bauri Cooli and Muslim *Mohalla*. In addition, youth do have some social interaction with their age mates from the Bauris and Muslim communities in a limited way. They are not very active in participating in others' social activities. The social interaction of Hadis reflects their illness behaviour and the process of seeking health care services from hospitals and doctors in the town. Choices of Hadis are not limited, but they choose to access treatment mostly in the clinics of doctors in their neighbourhood. Therefore, I have studied perceptions of illness among Hadis as separate from their social structure but focussed on the social interaction of Hadis, which is crucial to study their illness experiences.

This poses a problem for researchers in the field as collecting information on illnesses requires very close interaction with people's illness experiences encountered at hospitals, clinics, medical shops, and traditional healers. Illness is not only an individual's experience but also social and familial. Illness is negotiated at the household level and sometimes represented by individuals among Hadis in Hadi Cooli, the area where I conducted my fieldwork from the early months of 2008 to July 2013, but not continuously and consistently over this period.

There is a constant interaction between the Hadis, who feel unwell and discomfort, and the household members. The illness reporting and discussion within the household on the severity of illness and the consequent effects of such illness help to decide the further course of action. Some people also seek over-the-counter cures for illnesses perceived as minor or non-serious, such as to relieve pain and uneasy feelings, headaches, and body pain. A Hadi crosses many such layers involving local practitioners before it is reported to the doctor for treatment. This is discussed in the form of narratives and in-depth interviews of members of the Hadi Cooli across the chapters.

1.1 Theoretical Framework

1.1.1 Lay Understanding of Perception

The book takes the concept of perception to understand perceived notions about illness and health in a caste living in a specific urban locality. While interpreting the information to explain and understand the social situations and environments, an individual employs the technique of categorising, recognising, and construing her/his sensory information. The development of perception passes through stages and schemes as mentioned in earlier sentences. The mind and thoughts shape perceptions, such as learning, memory, expectation, and attention (Gregory 1987). In this process, perception differs from one person to another, and consequently, people perceive one situation quite differently. People from Hadi Cooli perceive the process of illness experiences differently from others due to their own individual and social background. Hadis, in the process of perceiving the world outside, also unknowingly begin constructing their perception by doing everyday work in their lifeworld (Husserl 1970). Hadis, as an individual from a lower caste, infer and develop their own perception based on their experiences.

Perception significantly differs from one individual to another, one group to another and one community to another but one also finds certain commonalities with castes and communities in few parameters. Experiences during social interaction and especially while accessing health care services in clinics and hospitals, there are a few instances when a patient develops some understanding in mind and at a sensory level and tries to give meaning during social interaction, especially while accessing health services in clinics and hospitals. By giving meaning to the information and experiences of a person, there is a possibility that perception would differ from one person to another according to the information received and interaction sought. For example, Hadis have their own perceptions about the health care system, hospitals, doctors, health personnel, and the health care providers in their locality and town or Block Development Office (BDO). Their perceptions are built in a process over the period. It is a reality that perception is essential in every human mind and society in order to understand each other and the objects in the environment.

Richard Gregory is of the view that there is a certain relationship of perception that results in the formation of 'Illness perceptions which are organised cognitive representations or beliefs that patients have about their illness' (As cited in Petrie et al. 2007, pp. 2–3).

1.1.2 Doctor-Patient Relationship

The Hadi relationship is also governed beyond the doctor-patient relationship in a clinical set-up. Doctors and medical shop owners also seek Hadis' help at a time when their domestic drains, pipes, and toilets are blocked.[2] This impacts the interaction with doctors while Hadis seek medical services and care. The individual's perception about himself or herself impacts the formation of illness perceptions; the representations and sometimes beliefs of the patients about their own illnesses are influenced by the perceptions framed through day-to-day interaction. In fact, the individual's perception may not necessarily be correct and true in the framework of the bio-medical model of illness and health. But there are chances that people are rational and logical in perceiving their own subjective notion of illness and health. People cope in illness and health problems in their own ways and normalise the bodily discomfort themselves, at home, and at the community level. People do provide a true sense of the problems felt and perceived in the body that shape future health-seeking behaviour and probable ways of coping with illness (Benyamini 2011).

The research illustrates various forms of perceptions in the process of interacting and encountering several institutions, namely health care providers. Studying illness experiences and perceptions is inter-connected and provides meaning and expressions in the form of perceived notions among Hadis. Hadis's experience in the process of seeking treatment required intense qualitative study to record various stages of illness and cure. Such perceptions of illness and health that are true to the people and may not necessarily be based on medical understanding are important to this study and core to the thesis.

1.1.3 Illness and Health

Illness and health have been represented by various measures, including physiological markers, self-perceived well-being, and reported morbidity (Aiach and Curtis

[2] Many a time relationship continues for longer period of time and also passes from one generation to another. Jamal Hadi, father of Gulrez Hadi worked till 1980s for Zamindar Tukan Mohammad, without asking for cash payment. But this relationship did not continue in the present generation. The Hadis are no longer dependent upon individual households and families. The client-patron relationship of Hadis also got dissociated when the son of Jamal Hadi found employment at the Bokaro Steel Plant (BSL). Such data helps to contextualise the social relationship of Hadis outside the Hadi Cooli.

1990; Bowling 1991). But the perception and reporting of morbidity are subject to what has been termed the 'iceberg' phenomenon (Last 1963; Wale 1976; Hannay 2011). This refers to a large proportion of symptoms and diseases which are not reported to medical practitioners but are found among people. McAteer, Elliot, and Hannaford, in a study on urban and rural populations in the UK comprising a sample of 2474, argue that,

> The symptom iceberg describes that most symptoms are managed in the community without people seeking professional health care. The size of the iceberg for many symptoms is unknown, as there is an association with personal characteristics, including a history of chronic disease (McAteer et al. 2011, p. 1).

This could include certain symptoms that are considered common in daily life, such as tiredness, joint pain, and headache, which do not sustain for a longer period.

Mechanic and Volkart's definition of illness behaviour is important for examining the concept of illness. They suggest, 'To the ways in which given symptoms may be differentially perceived, evaluated and acted upon(or not acted upon) by different kinds of person' (Mechanic and Volkart 1960, p. 87). In their paper 'Illness Behaviour and Medical Diagnoses', they write that the consideration of seeking health care also depends upon the class to which an individual belongs. The study was conducted on 614 male students in a university. The students identified and prepared a matrix of 16 categories that made them decide whether to approach a physician for diagnosis. They further refer to a study conducted by Koos (1960) that identifies different illness-seeking behaviours depending on socio-economic status. Patients from lower socio-economic conditions are more likely to ignore their illness than those from the upper financial strata. In view of this explanation, the definition of illness behaviour relies upon how patients report their symptoms later to evaluate the state of illness and follow up to report to the physician. It is a significant fact that 'The variables affecting illness behaviour come into play prior to formal medical processes that have been initiated. In this sense, illness behaviour even determines whether diagnosis and treatment will begin at all' (Mechanic and Volkart 1960, p. 87).

A study by Morrell and Wale (1976, pp. 398–399), in which 198 women recorded their daily health in diaries, showed that 57% of women (20–44 years) did not report minor illnesses to a doctor, showing a preference for self-care than seeking physicians' support. Factors such as education, religion, income and class membership, occupational status, and whether the illness affects everyday work affect the illness behaviour and health-seeking practices of people. Some patients respond to even minor pain and discomfort to quickly seek whatever medical care is available. People from low socio-economic backgrounds tend to delay in seeking health care services, which increases the illness level. McBroom (1970) writes that a satisfactory explanation proves a causal relationship between illness and the socio-economic status of the people. However, McBroom provides another explanation that those who belong to lower class status are biologically inferior. As a result, disease rates are higher in that strata. However, he also admits that this explanation is not accepted widely.

The level of attention given to certain symptoms if they affect everyday work is higher among people from lower socio-economic conditions. Meanwhile, their affluent counterparts are more likely to report their illness even if it affects their everyday work. Mechanic (1968, pp. 259–266) cautions that the illness rates may vary due to the population selected in the study, respondents' non-cooperation in providing answers (data), and problems in measuring the illness. The demand to work every day among economically weaker social groups is high as one cannot afford to lose wages for the day and simultaneously the cost of the treatment. Such external factors do influence illness behaviour of people.

Household-earning people cannot afford to be bedridden. Therefore, the constant effort of society and households to help a person come out of a 'sick role' (Parsons 1951) is an important aspect of illness. It helps individuals to perform everyday social and economic activities. The expression of symptoms of discomfort and pain demands from an individual depends on how one perceives those symptoms, the level of pain, and whether it would affect the work and routine activities or not. In the next stage, a person seeks household discussion regarding the perceived illness. These situations help decide the illness level and whether it is to be reported or delayed. The choices of healers are also dependent on the family's beliefs and traditional systems, etc. The mutual comfort in past encounters/illnesses helps to decide the doctors to be consulted. Such experiences do not necessarily have to be of the individual but that of members in the household and, sometimes, members close to the person within the community. From the above discussion, it can be surmised that illness and illness behaviour need wider discussion on various factors.

1.1.4 Healers as Health Care Providers

Reporting to healers helps decide whether the person's social role is 'sick role' or not. If it is, then the cultural expectations surrounding the sick role, the structure of lay referral systems, sub-cultural expectations of health, and the interaction of symptoms with the demands of a person's life do help shape an individual's illness and illness behaviour (Mechanic 1968).

Self-reporting of symptoms shows how an individual perceives the symptoms of bodily discomforts; they are evaluated to identify the further course of action (Mechanic and Volkart 1960, pp. 87–88). This involves people answering questions, and their responses show how much they are influenced by their own perception of bodily state. The self-perceived notions of individual health help interpret the nature of reported illness. These are also shaped by the quality of locally available medical care, and the type of doctor-patient relationship which can sustain the interpersonal relationship between medical care, doctor, and patients (Blane et al. 1996).

The extent to which an individual is reported to be unwell after diagnosis enters into a different social role as sick. In such situations, illness is found to interfere with the performance of everyday social roles and disturb an individual's daily activities, whereas non-performance of daily work is seen as a deviation and is considered problematic for people in society (Blane et al. 1996).

Illnesses often interfere with performing normal social roles and even economic activities. Those individuals who are unable to perform and attend at workplace, take care of dependent members, get disapproval in society. Rather, it is expected that such individuals return to normal life as soon as possible and perform routine activities. The study's population tends to avoid minor pain and bodily discomfort in view of the fact that such roles restrict them from going to work, and consequently, there is a loss of earnings. The lower class such as Hadi population does exhibit this above explanation. Kari Devi narrated, '*Hamni ke kaam hai tohe kalwa-biyaari hau, hotelwa mein bimaari nahee chaltau, malikvan ke ye sab baat naae jamtau, ke hamae bimar ho galiye.*' Literally, we do not have rest from work; only when we work can we get food. Working at hotels is difficult, and the owners do not understand and accept the excuse of illness. The hotel owner, my employer, needs his work to be done every day.

The study would help in understanding the health care and illness behaviour among Hadi workers (sweepers) in Chas and the ways in which illnesses are ignored, delayed, and consciously avoided, thereby not entering into the sick role unless it becomes a major illness. A conscious attempt is made to delay and not to affect everyday wage earnings that go into fulfilling the basic needs of the households. I argue that there is a process and a long period taken by an individual before sitting idle to rest during illness. The Hadi workers are found to be quite aware and careful not to fall ill and take rest from work. The general perception among the workers does not allow them to accept the sick role unless it is beyond their control. In this context, Chap. 4 on 'Perceiving health care services through suya-goli' at davaee dukaan and private clinic is relevant to examine the health-seeking behaviour of individuals.

It is another matter that those individuals who report illness, seek health care, and are diagnosed unwell attract sympathy and support towards getting proper health care services in order to allow them to return to routine activity. It is also found that illness is a universal human experience which comes to be endowed and manifest with cultural meaning. The legitimate non-performance of normal social activities due to illness is one part of the 'sick role'. The advantage of legitimisation of the individual as a patient by medical professionals offers certain benefits, namely both absence at work and decreased physical activity at the household level. They receive familial and social support that helps them to come out of sick role performance into normal performance (Parsons 1951). The idea behind this point is to accentuate the process of how an individual reports illness later to be legitimatised as a sick role, a status that keeps an individual free from work and, in return, demands the attention of household members to provide care services. However, the pressures of fulfilling the household demands bring the patients back into regular action. It is in this context that illness is universal, but they do not have similar meanings and vary according to the references made by various social groups; Hadis, in my research, would not opt for a sick role until they are not able to manage illness.

Through this research, it is also found that health perceptions define how individuals perceive their health, whether they perceive that they are in a healthy state or an ill one. The decision to have an illness depends upon an individual's bodily symptoms, but it is also important to know whether that discomfort disrupts routine activities. The data finds that people keep delaying illness, reporting minor illnesses, and seeking temporary relief through over-the-counter drugs from medical shops. An attempt is made to continue the work schedule until asked to go on rest by the doctors. Doctors (Hadis refer to RMP practitioners as doctors in Chas) play a pivotal role, and a decision to rest is taken only when the doctor strictly advises it. However, visiting a doctor is a second stage in seeking health care during illness. The social process between *davaee dukaan* (medical shop) and RMP's clinic helps to examine the perceptions during illnesses.

Health perception differs among individuals. Why two individuals with the same level of general health have different health perceptions is to be understood through theoretical propositions based on the social classes of the populations, as argued by Blane et al. (1996). In their study, it was found that class differences in illness behaviour affect self-reported health; this was found among the Hadis between households having employment in steel plant, Nagarpalika workers, MADA employees, and others. The lack of knowledge, awful experiences in seeking health care, and the dominant factor of the occupational hazards associated with the menial job of manual scavenging emerge as important hurdles in the treatment-seeking process. Hadis do not narrate the minor problems that minimise their work capacity and obstruct their everyday routine because they do not find it important to report them. The lack of understanding of illness symptoms, limited social interaction with people outside the caste group, and the inability to access quality public health care centres are some of the chief reasons that result in ignorance or lack of reporting of illness.

Ramesh Hadi, in an interview said, 'It is quite easy to avail the services for employees at the Bokaro General Hospital, but one faces several difficulties in accessing Chas Hospital. If you know someone in Chas Hospital, then things are quite smooth and easy, but none of us have got any contact in the hospital.' This also explains the reliance on medical shops [*davaee dukaan*] and RMP clinics. Such resources are available to them in their vicinity, unlike the hospital located far away.

The perceptions of illness are studied based on reported illness among the various social groups. The questions to report illness of the study's populations are addressed in my study among the Hadis engaged in menial works. Health perceptions are important because they influence the individual's ability to participate in various physical, social, and occupational activities. Once an individual perceives ill health, it is likely that his or her previous level of activity will decline (Julia et al. 1989).

For these reasons, an awareness of social situations, the availability of various institutions in their cultural context, and health values are essential in order to understand the patterns of individual behaviour in interaction with the environment. Pelto and Pelto (1997) understood health beliefs and health behaviours are a subset of this relationship that takes place between factors influencing illness behaviour in the process of reporting illnesses and in seeking treatment.

Individuals inherit from their culture's structured vocabularies of health and illness that limit possibilities for interpreting physical and psychological states (Dimiuou 1995). The meaning associated with certain vocabularies used in narrating illness symptoms and discomforts is culturally structured. The words chosen to describe a few bodily dysfunctions do not mean the same thing to a Hadi as an outsider.

Tatainee (a constant pain in entire body), *Adhkapari* (headache in half portion of head), *karjawa-faet jaa hau* (extreme pain and suffering), and *gaud-haath peera haua re* (legs and arms in pain) are a few words and phrases that show the Hadis' way of expressing illness and discomfort. Local healers, RMPs, and druggists at medical shops, however, can understand the underlying meaning and provide symptomatic treatment according to the illnesses stated. The health-seeking experiences bring certain colloquial words into the illness narratives of Hadis that are ignored in most cases in doctor-patient relationships. Another aspect is that each individual and household would have its own way of narrating the illness-related problems to the doctor.

1.1.5 Traditional Healing Practices

It is important to see that traditional beliefs about illness and health problems are a different and difficult area of research to study, especially correlating it to scientific theories of illness. Over the period, the advent of bio-medical medicines has influenced traditional healing systems, but the changes brought do vary. The Hadis, as a community, believe in seeking care through traditional healing methods, but there is a decline in the practice of traditional healers in their Cooli. Similarly, non-allopathic medicines are also used less often, although they still retain their popularity.[3] However, the popularity of Ayurveda has gone down, but Ayurvedic prescriptions are still found, probably due to the fact that despite the change, people always rely on traditional means that have been tested for generations in each society for managing illnesses. For instance, the application of turmeric to small cuts and injuries has been found for generations, and the use of lukewarm water to relieve throat irritation is a common practice in most households. These are some of the household healing practices.

Traditional healing also includes incumbent rituals and beliefs that are supposed to provide cure and prevent illness. For example, in the case of chicken pox, it is still traditionally believed that there is no need to seek treatment. Hadis do not refer to persons infected with chicken pox as ill, rather they believe it to be by the grace of '*Sitala Mata*'. This belief is also found among other Dalit communities such as the Bauris in Chas; there is even a *Sitala Mata* temple at Bauri Cooli. Elders in both communities (Hadis and Bauris) report that smallpox was a threat to the lives of Hadis and Bauris before the 1980s in Bokaro. The presence of *Sitala Mata*, a deity, is found to be at pan India level. Eastern states of India have the important role in Sitala and reporting of smallpox, as widely discussed in two of the studies by Nicholas (1981) and Misra (1969). Though there is less reporting in the present scenario, it

---

[3] For instance, the practice of Ayurvedic medicine is still popular as evident from the clinic of Ayurvedic doctors who practice at their own medical shops located in the Main Road Market in Chas. There is perceptibly a different class of patients who seek treatment from Ayurvedic doctors. The cost of treatment in Ayurveda is beyond the paying capcity of Hadis.

is said that no one seeks medical intervention during the arrival of *Sitala Mata* in the human body. However, doctors point to the fact that there are a few medicines that can be prescribed to patients with smallpox. It is a kind of viral infection and anti-viral tablets help to regain normal health in about a week.

Dr. Vimal Singh, in one discussion, said, 'It is not uncommon to rely upon *Sitala Mata* and other traditional beliefs. My parents also believed in this Goddess. People do not know the reasons behind smallpox and how to cure this disease. A few years back, we had no medicines to prescribe, but now anti-viral medicines are available. Though doctors don't prescribe allopathic medicines, patients are believed to recover on their own without medical intervention. Even I think each individual can fight against such viral infections unless the patients are quite weak and dull.' This data testifies to the gradual change in people's belief system in the field.

To substantiate this, one can see the narrative of Ramesh Hadi, 'It was never acceptable to us (Hadis) to seek medical intervention for smallpox and even if someone was mentally disturbed. I could understand that both these are also a kind of illness only after I was posted in the mentally retarded ward in Bokaro General Hospital (BGH). It was quite difficult and unravelling, and all traditional beliefs and understandings were challenged. Had I not been posted in the mentally retarded section, I could not have had a chance to listen to the discussion and see the treatment process when doctors arrived in the ward. The discussion between doctors and nurses was more important and meaningful to me.' This narrative is interesting as it shows how the traditional form of medicine is gradually being replaced by biomedicine. Traditional healing practices have been losing ground in the strong presence of a bio-medical system of treatment. For instance, Kari Devi said, 'A tablet brings more relief than any other method in body aches. Massaging my body in severe body pains demands the presence and help of another person at home whereas I get a few *goli* (tablet) at *davaee dukaan* that provides relief in sometime. In case of severe and prolonged problems, Dr. Vimal Singh provides *suya*.'

There is a gradual strengthening of scientific theories of illness, which usually employ biological explanations for disability or illness (bio-medical model, microbes, etc.) and emphasise disorganisation or dysfunction at a cellular or inorganic level (Haron et al. 2004, p. 60), rather than the traditional. With the gradual transformation of the population from rural to urban, there is greater reliance on the bio-medical model; therefore, allopathic medicine rather than traditional healers. The variation and battle to mobilise the cost of treatment and seeking cure during illnesses are managed according to the available health care systems and practices of the community.

One also examines briefly how health practices depend upon generational references: on the one hand, there are many beliefs, customs, and attitudes that are transmitted from one generation to another. There is also reliance on the older generation for advice. For instance, a few youth stay connected to Ramesh Hadi, Somesh Hadi (Somesh is an employee at BSL), and others for advice on social and familial matters. Ramesh Hadi says, 'Now people don't want social interference in their family affairs, but when there is any problem, they look out for us. We, as elders, want to transmit the knowledge of our Budha-Purna (ancestral wisdom) to our children so that they understand the history and culture of the Hadis.

Many cultures visualise illness as a battle in which there is a possibility of success and failure, and the exercise of control over the illness demands certain different factors in favour of the sick patients (Seale and Jonathan 2008). Illness behaviour is not only controlled by cultural factors of the community but also by individuals' perceptions that are constructed in due course of suffering and experiences in seeking health care. There is a point when a patient also loses hope in getting treatment as a result of which there are feelings of hopelessness and powerlessness. There is no control over health care availability vis-a-vis the constant struggle and attempt to delay and ignore bodily discomfort. The delay in seeking health care is also economic and cultural, as Hadis feel they should not take off from work as much as possible. Ramesh once clarified that, 'I never took leave on the pretext of being ill until I was strongly advised to take rest or I had no control over my body. It is never a choice but our job is also extremely crucial in all public and private organisations. Our supervisor does not want the sweeper to take leave. It directly affects the office. There are barely any extra staff members that can do extra jobs in absence of sweeper.'

Secondly, the expression of illness is gendered among the Hadis, as found in some other studies (for instance, Seale and Charteris-Black 2008). Kaliya's wife Puja Devi, said, 'I never want Kaliya Hadi to fall sick. I also stay cautious and do not allow him to delay if he has some discomfort and feelings of being unwell. It is essential to my household that my husband stays active and healthy. I ignore illnesses of mine but do take care to consult Dr Vimal Singh in case of his illness and both of my children." However, in an interview, Kaliya narrated, 'I do not have much understanding of hospital as an institution and what kind of illness is threatening us. It is difficult to have command over illness and disease, hospital and its functioning." Doctors' words and explanations are relied upon. Going to Suhail, the man at the Chas medical store, to take pills for minor illnesses (body aches, headaches, mild fever, stomach disturbances) is a regular practice. Hadis are not perceptive to different modes of treatment that ought to be given to males and females, and, as a matter of fact, children. It is generally discussed in the household and then illness is evaluated in terms of severity and discomfort. Women are found to give preference to their husbands and children more than themselves, and this inclination and feeling are also reported by their husbands. It helps to analyse that illnesses have different connotations for different genders. Hadi women may feel body ache and face reproductive problems but do not report it to a doctor.

Many studies have alluded to the disruption caused by disease in daily life, though the degree of influence may vary. As Michael Bury says, chronic illness presents a disruption and uncertainty in the everyday lives of people: The patient's structure and knowledge of everyday life including taken-for-granted assumptions are disrupted as they are forced to pay conscious attention to their body, be aware of the pain, suffering and potential death (Bury 1982). 'The advent of the illness is disrupting what people felt was normal, disrupting a future biography, and challenging their everyday physical, emotional, and social lives' (Faircloth et al. 2004, p. 243). Bodily disruptions also occur to a person's sense of self; it forces them to renegotiate their personal and social networks (Kelly and Field 1996). These disruptions to social networks and community involvement are due to functional limitations, the inability to maintain normal activities, embarrassment, and stigma brought on by the illness (Bury 1982). In response to these disruptions, the patients rearrange their social network by mobilising new resources, which provide new support and aid in the course of the illness (Bury 1982).

Ruby Bhardwaj, in her paper 'Medical Pluralism in India: The Interface of Complementary and Alternative Therapies with Allopathy', discusses how patients in her field were getting disenchantment and disillusioned from allopathy treatment. Out of 108 patients, 53 patients were found to be having disenchantment from allopathy despite the fact that they sought treatment and help from highly specialised allopathic doctors and big hospitals. They tried to seek better treatment, but nowhere the sense of satisfaction and relief from their illness were provided and found by patients. It was a kind of forced outing from biomedicine towards complementary treatments or alternative medicines. In fact, one-third of the respondents reported having side effects of the allopathic treatment. For instance, a few patients suffered arthritis, one of the lifelong perceived illnesses, due to the consumption of steroids and 'This iatrogenic condition was beyond the scope of effective allopathic treatment' (Bhardwaj 2010, p. 35). Similarly, obesity was found to be a common side effect of allopathic treatment. Ruby Bhardwaj gives an important account of how patients were aware of the side effects of allopathic and one-third of her respondents wanted to stay away from this medicine due to the *garmi* (heat) allopathic medicine creates in the body of patients.

Hadis in Chas also keep moving from allopathic to alternative medicines. They have been aware of the available alternative and traditional systems of treatment for years, much before they became dependent on allopathic treatment. Perhaps the approach of allopathy is limited to symptomatic treatment and medicines are prescribed accordingly. This method does not necessarily address the satisfaction of the patients. In Bhardwaj's study, respondents also believed that allopathy can be treated symptomatically but cannot eradicate the disease from the body. Hadis have chosen traditional healings, approaching *Baba* to find comfort and solace in illness. They also continue buying medicines over the counter and visiting their doctor's clinic in the locality. Bhardwaj included practitioners in her study as complementary therapy (CT) who also had formal training in allopathy, naturopathy, and homoeopathy complementary therapy (CT) who also had formal training in allopathy, naturopathy, homoeopathy, and physiotherapy. In her work, Bhardwaj (2010, p. 206) has used Nichter's term *Masala Medicine* briefs the way practitioners use all sorts of medicines and alternative therapy. Masala medicine is practised in one room, combining Ayurveda and allopathy to provide quick relief to the patients. It appears that *Masala Medicine* is significantly not coming up in the present study population and field area, but certain reflections can be traced in the discussion provided later in the chapters.

1.1.6 Health and Health Care Services

The debate on illness and health is interdependent, and both find a close association with the notion of disease. Frank (1997) wrote that 'Disease is not the absence of health; rather, health and illness are the complementary cycles of the world' (cited from Radley et al. 1997, p. 398), whereas 'Health has been conceptualised as a state, a process of development, an actualisation, an outcome, and a style of life' (Meleis 1990, p. 108).

Smith (1981) divided health definitions into four broad categories: clinical, which represents the biological model of disease; role, which addresses an individual's ability to meet societal expectations; adaptive, which focuses on ecological explanations; and eudaemonistic, which focuses on high-level wellness (eudaemonistic refers to a philosophical doctrine that holds a view in the value of moral action that can produce happiness and joy).

Health status and perceptions are two important crucial concepts to research into health construct. Health status, the more concrete of the two, is consistent with Smith's (1981) clinical, adaptive, and role performance definitions. Epidemiological and bio-medical research often report health status findings regarding risk factors likely to lead to increased morbidity and mortality. In contrast, health perceptions are more challenging to conceptualise. Perceptions often rest at the unconscious level and are shaped by personal experiences and expectations. Webster (1993, p. 1675) Third New International Dictionary defines perception as 'The integration of sensory impressions of events in the external world by a conscious organism, especially as a function of no conscious expectations derived from experience and serving as a basis for or as verified by further meaningful motivated action.'

It is likely, therefore, that one's health perceptions rest below the surface, are moulded by experiences, and are reflected in actions. To develop an understanding of health perceptions, a valuable definition of health directs the researcher to the sets of activities and actions that affect health and reflect health perceptions. For these reasons, operational definitions of health that solicit cultural understandings of health and contextualise the cultural aspects of life into the definition are important. All health definitions are limited because none encompass the complete range of health experiences for all individuals; however, the World Health Organisation's (WHO 1978) definition of health as 'A state of complete physical, mental, and social well-being and not merely the absence of disease and infirmity' is valuable because of its inclusiveness. This perspective 'Weaves health through the fabric of everyday life in which activities that support and detract from mental and social well-being are considered alongside the physical elements of health' (Mendelson 2002, p. 13).

However, contemporary health care places great emphasis on identifying, responding to, and modifying risk factors for disease. There is a large body of literature on patients' or laypersons' understanding of risk (Parsons and Atkinson 1993; Scott et al. 2005). However, my analysis does not focus on risk but is limited to illness and disease. It is pertinent to note that the differences in the health care services and their availability are significant in understanding the perception of the deprived communities.

Stratification based on inequality and unequal allocation of health care services is reflected in the scarcity of facilities; who are privileged to control resources and facilities, in contrast to the classes at the bottom, who have no control and no access, are deprived of even the bare minimum necessities of life, not just health care needs. This is true whether it is health, housing, sanitation, educational facilities, or employment, as shown by Djurfeldt and Lindberg (1980), Banerji (1981a, b, c), Zurbrigg (1984), and Qadeer (1995). Though these studies focus on the rural stratification and the poor, the case of the urban Dalits like Hadis reflects similar conditions.

The social inequality of Hadis has serious implications on their health, not only at the workplace but also at home. This is because skills and education and hence job definitions of the workers depend on the socio-economic background they come from. The more depressed their origins, the more likely they are to take unskilled jobs and jobs that no one else takes because of the associated hazards. This becomes self-perpetuating because the less ability they can resist these jobs, the more disadvantaged they will become by constant exposure to health hazards and the ensuing illnesses and injuries (Qadeer and Dunu 1989). However, despite these challenges, the Hadis community at Chas town stands as a testament to resilience, navigating through the social hierarchy to access basic amenities for life like housing, drinking water, medical facilities, education, etc.

1.1.7 Health Care Utilisation

DiMatteo and Freidman (1982) identify three factors which may influence the decision to utilise health care services: first, the background of the patient; second, the patient's perceptions of illness; and last, the social situation. Social situation refers to all the social factors that influence an individual. Irving K. Zola, in his paper 'Pathways to the Doctor-From Person to Patient[4]' identified five social triggers that influence the judgement of the patients to seek health care service needed to manage the illness. They are: first, perceived interference with physical activity (work related); second, perceived interference with social or personal relations; third, an interpersonal crisis; fourth, setting the deadline (if I am not fine by Friday, I will consult a doctor); and finally, pressure from family and friends. These triggers are also relevant when approaching the health-seeking practices of the Hadis in their social context (Zola 1973).

At times, people perceive and do relate religious beliefs with illness experience that disease and illness are caused by God, as found by Pargament et al. (2004) in their study of 268 medically ill patients. In the two and half years follow-up study on hospitalised patients after their discharge from the hospital, they found that religious

[4] Irving K Zola studied 400 adults with disability in a village in Netherlands Het Drop provided different triggers that influence the patients to seek treatment.

beliefs are a coping mechanism during sickness. Hadis have sometimes adopted religious and spiritual methods to cure the illness and find healing. The utilisation of health care services became an important factor to observe and understand the perceptions of illness among Hadis in the process of seeking health care services.

I have used a few components of Anderson (1995) Health Care Utilisation Model, which provides special insight by predicting health care and seeing the levels of utilisation. The model helps to understand the behaviour of patients and the patterns of utilisation. The model provides three different categories of variables that influence the use of health care services and those variables are distinguished as: need factors, enabling factors, and predisposing factors. Firstly, predisposing factors include demographic factors, social structure, and health beliefs. Demographic factors are age, sex, sex ratio, and reproductive status. Educational level, ethnicity, race, occupation, family, religion, etc. are essential components of social structure.

The enabling resources include easily available health services and health organisations (hospitals, clinics, etc.) that manage and support health care delivery as service providers, health policy, a social network of patients, and infrastructure to provide health care.

I use the component of demographic characteristics in Chap. 2 to illustrate the population structure, age, and sex composition of Hadis in my study area. The second component, enabling resources, helps to contextualise available health care services and the pattern of health care utilisation by the Hadis in government hospitals, private clinics, and hospitals. Anderson's model for understanding health care services and utilisation and how my study population perceives their illness become essential in the analysis of their utilisation of health care and experienced health care and impediments to it. I studied the various forms of coping mechanisms as practised among Hadis. The entire process is debated at various levels across the thesis chapters. In addition to the factors discussed in Anderson's model, I also discuss caste-based traditional occupation as an important factor in deciding whether the person in illness access health care. Caste-based occupations of sweeping (earlier manual scavenging) do affect thoughts and behaviour while seeking cure Caste-based occupations sweeping (earlier manual scavenging) do affect the thoughts and behaviour while seeking cures in hospitals and clinics. It is perceived that associating with a menial and lowly occupation has kept them at the lowest strata in society. Given the model, my study does not utilise the complete model of Anderson. I have partially found it helpful in explaining the Hadis, quite a small population compared to the macro or vast population of Anderson's health care utilisation model.

1.1.8 Occupation, Caste, and Illness Experiences

It is vital to include caste and occupation in understanding the perception of illness and illness experiences. Massey (1997) argues that the process of internalising an event is integral to human behaviour and, in turn, influences life activity and chances. It is pretty obvious that the myriad forms of discrimination due to caste and occupation (Khare 1984) shape the mindset. Discrimination is not only about the occupation. Rather, this area of study, largely referred to as the Sociology of Dalits itself, is neglected in Indian Sociology, similar to what other marginalised groups argued (Kumar 2005). Hadis perceive themselves as 'unclean' because they perceive their work as menial and also because of the caste to which they belong. They are aware of their filthy working conditions. They also work in an unhygienic environment without gloves, and it was observed that most do not even wear footwear at work. They clear up manholes when clogged, exposing them to several kinds of infection. The hazards include exposure to an extraordinarily harmful and complex mixture of toxic gases that may contain methane, carbon dioxide, sulphur dioxide, nitrous oxides, and hydrogen sulphide. Hadis did not use the term hazardous chemicals term' hazardous chemicals, but they explained it in discussion in different words. Anup said, 'It smells awful. I would sometimes vomit after cleaning a manhole or drain due to severe bad smells and gases. Vomiting is a common problem due to such work. Water clogging and blocked drains are mostly during monsoons blocking rainwater from flowing, blocking rainwater from flowing. However, it is also a work round the year.' The relief from cleaning drains demands every day is due to improved sewage systems and less frequent problems of blockage of drains that demand entering into manholes. Mostly it can be cleaned while entering half into the drain. Hadis also felt thankful to God that the entire town did not have a closed drain system.

Rajinder Hadi once said, 'The dirt and filth are always menial. We get cuts and bruises on our hands and feet quite commonly. The sewers carry chlorine bleaches, household and industrial waste, and human excreta. Chas has also had small industries in the recent past that were not the scene there ten years back. You can see the waste of unused diesel and mobil (disposed engine oil) into the drain. Sometimes drain water is pure black.' The challenge to clean the drains has increased over the years with the growth of the town. This caste-based employment has also been found to be associated with higher chances of health hazards.

A few related studies focus on poverty and recognise poverty as an essential determinant of access to public health care services (Carstairs 1955; Zurbrigg 1984). The curse of poverty on a few caste groups and the rather unequal distribution and allocation of resources in caste-based society are a critical reason to mention. As mentioned above, a few studies by Banerji (1981a, b, c), Qadeer (1995), Acharya (2011), Acharya et al. (2015), Ziyauddin (2022), and Sahu (1986) find that income inequality is of prime importance in unequal accessibility of health care services especially public health services in rural India.

It is important to mention private and public health care services at this point as it strengthens my point that the perception of Hadis is interwoven with their experiences while interacting with various social and health care institutions. The simple correlation between low-caste status, poverty, and health care inaccessibility is inconclusive. The role of public health care services/centres is significant to elaborate on the perception of illness and health among Hadis. The availability and accessibility of health care centres in terms of physical accessibility, socially accessible accessibility, or social accessibility are important dimensions addressed in the research.

The stratified social system (Khare 1984; Shah 2002) operates through various institutional mechanisms and processes (Patel 1994, 2008). The relationship between health care and the kind of experiences witnessed among the Hadis is intertwined. The intention is to seek elaboration on the perception of illness and health where the utilisation of health care services is an important factor to be mentioned.

Several studies (Acharya 2010a, b; IIPS and ORC Macro 1995, 2000, 2007) have shown a relationship between the low-caste groups, i.e. Dalits and low levels of health care utilisation as compared to others (Ramaswamy 2005; Acharya 2002; Kulkarni and Baraik 2003; Srinivasan and Mohanty 2008). Thorat and Sabharwal (2006) have also shown close links between low utilisation of health services by Dalits and their high poverty and illiteracy levels. Apart from the above-mentioned studies, other studies on utilisation of health services have not paid primary attention to the above-mentioned studies, and other studies on the utilisation of health services have not paid primary attention to the perception of illness and health among Dalits, especially in policy formulations.

Acharya (2010a, b), on the window counter, based on her field data in Western Rajasthan, focussed on discrimination in rural public health care services; it helps to frame one aspect of the current study. Such instances include less attention, an unsympathetic attitude, and rude behaviour towards Dalit patients. Pharmacists are also known not to dispense medicines properly, often leaving them on the window counter without explaining the medicine doses properly. A similar study on Bauris and Hadis patients in Jharkhand demonstrated how their caste acted as a barrier to accessing health care services and burdened households by spending a higher amount on their illnesses and disease (Ziyauddin 2022).

The formulation of the research questions has to take the dimensions mentioned above into account and also count for the probability of an upper-caste lens, including bias and prejudice towards the living and social situation of the Hadis. It is primarily essential to see how Hadis in Chas (district Bokaro) return to their regular routine activity after getting health care services. A member often does not want to be caught in an illness cycle. However, if caught, the proper treatment of illness from an objective medical approach helps the patient recover quicker. The decision to seek health care from a professional would depend on the socio-cultural context and network of social relationships in which an individual resides, as discussed earlier (Weiss and Lonnquist 1997, p. 135). This socio-cultural context is well-appreciated by the doctors (RMPs) in the locality of Hadi Cooli. The narrative provided a framework to analyse the data obtained.

1.2 How the Intensive Study Was Used as an Ethnographic Methodology

The study's objective was to examine and analyse the perceptions of illness and health among the Hadis in Bokaro. The perceptions of illness and health are examined particularly in Chaps. 3, 4, and 5, guided by the data collected in the field. Hadis form a different social group among the Dalits in India, with peculiar socio-economic conditions and occupational associations that make this study unique and pioneering.

The present study is based in the state of Jharkhand, where the Hadis reside. This was briefly discussed at the beginning of the introduction, and a detailed analysis is given in Chap. 2 to comprehensively understand the health status of Hadis and explore their perceptions of illness vis-à-vis health and their access to health care.

For the purpose of research guided by the above-mentioned objectives, I selected an urban settlement, Hadi Cooli, a Scheduled Caste Cooli (Mohalla), in Chas town in Bokaro district. Hadi Cooli is one of the SCs' populated Cooli, with 51 households (a detailed description is provided in Chap. 2). I used narrative to examine the illness experiences of Hadis.

1.2.1 Narrative Frameworks

The question of how one understands the perception of illness and health depends upon social context: community, institutions, and regions of residence. Narratives offer insight into social contexts as sites for the construction of meaning (Mumby 1993). The implication of social roles in creating meaning in a particular social context is an important element in the study of narratives about the experiences of health and illness. A narrative can be explained or understood to organise a sequence of events into the whole so that the significance of each event can be understood through its relation to the whole (Elliot 2005). In other words, narrative helped in my research to convey the meaning of events.

Mishler (1986), in her volume, acknowledges the need to listen to individual stories in the context of qualitative interviewing and brings in various processes (see Patel 1994) of illness suffering. Elliot (2005) mentions the growing number of publications on narratives that capture patients' illness experiences and perceptions of suffering through stories and narratives. Narratives as an inquiry provide great help in collecting data on ill health and individuals, which was quite helpful in my data collection (Kleinman 1988; Charmaz 1999).

It empowered the respondents and researcher to unravel those experiences that were not reported in the general discussion. First, it also paid attention to respondents' accounts by eliciting an interest in people's lived experiences and an appreciation of the temporal nature of those experiences. Second, a desire to empower research participants and allow them to contribute to determining what are the most salient themes in an area of research. Thirdly, an interest in the process and change over time; fourth, an interest in the self and representations of self; and last, an awareness that the researcher him/herself is also a narrator (Frank 1994, 1998). In Labove and Waletzky's (1967, p. 12) words, the narrative provided me with 'A method of recapitulating past experiences by matching a verbal sequence of clauses to the sequence of events that eventually occurred.' The data on the social history of the study population has meaningfully helped to fill the dearth of available studies on them.

How do people evaluate and make meaning of their experiences of illness? The range of evaluations provides emancipatory and enabling devices with which sufferers or their loved ones are seen to rationalise, the experience of illness and attribute meaning to it. Sometimes, the illness's meaning is unclear, even for the sufferers and close family. Illness narratives make connections with the social world in which they exist and provide validation for the experience of illness. This is in accordance with Frank (1994, p. 13) claim that 'Illness is a form of absence from the social presence that health takes for granted.' Absence does not imply the absence of illness in the human body. However, the ways in which illnesses render sufferers absent from social life and the ways in which illness are socially constructed and narrated. I have studied illness socially constructed by people in the field as '…Illness narratives are not an illness' (Frank 1994, p. 41), but they represent forms of illness.

Arthur W. Frank has argued that people suffering from profound illness tend to tell three predominant stories; he defines profound illness as that which is 'Perceived as lasting, as virtually affecting all life choices and decisions, and as altering identity (Frank 1998 p. 197).' The three predominant stories include;

(a) The restitution narrative,
(b) The chaos narrative, and,
(c) The quest narrative.

The restitution narrative exists because of the relative efficacy of medical professionals in diagnosing and treating physical illnesses adequately. As told, this story follows a general linear trajectory, in which an individual becomes ill but regains health through medical intervention and support of health care systems. This is a popular and pervasive account of illness. Most people assume that health will be restored after a certain period and with the necessary health care services. When medical interventions fail, the person might feel a sense of injustice or disbelief, leaving him or her angry, in despair, or in disbelief. The presence of the belief that a restoration to health will ultimately occur in the future and that it is simply a matter of time before that takes place can follow these feelings. Some people adhere to the hope of restoration until they die (Frank 1998).

In contrast, the chaos narrative is filled with uncertainty, confusion, and a sense of powerlessness over the events and illnesses in one's life. It refers to the perception that everything is falling to pieces and out of control. For the listener, this can be the most difficult account for which to hear and record the data. Frank (1998) has described this type of story as suffocating. For this reason, the chaos narrative is likely to be weakened, if not by the person telling it, then by the person listening to it. A common thread in this narrative is a sense of never-ending suffering and a series of negative or adverse life events that either contribute to the experience of the illness or are exacerbated by its presence. Frank (1995) argued that listeners are most frightened of the chaos narrative because it forces them to recognise their vulnerability to illness. It is unnerving and isolating to consider that one might neither regain health nor become spiritually transformed by the illness experience.

The last transformation through illness is evident in the quest narrative. In this account, a person doesn't have to return to health as it is in the restitution narrative. Instead, in the quest narrative, a transformation occurs. The emergence of an optimal state compensates for a failure to return to one's previous state. For example, a person with HIV disease might suffer and eventually die. However, after his/her HIV-positive diagnosis, the person becomes an advocate for those who are also afflicted with the disease. His/her transformation does not result in a return to physical health. However, it helps the person to gain emotional clarity, and, for the first time, a more profound sense of meaning to life may emerge in the thought. Although listeners to illness narratives might be most comfortable with restitution stories, in the absence of such accounts, they might hope for quest narratives. Sometimes, the quest narrative is encouraged, hoping to reflect part of the journey en route to restitution (Frank 1995). I find such a categorisation important in approaching narratives of the Hadi Cooli. Diane Lewis notes, 'Anthropologists who study their own societies will also add immeasurably to their theoretical understanding of mankind' (1973, p. 590 cited from Unni: 61). I conducted fieldwork in my region with the familiarity of residents and localities.

As part of intensive fieldwork, narratives helped me as a device facilitating empathy. It provided a form of communication in which the respondents could externalise their feelings and the most significant experiences. Similarly, the ability of an individual to interpret the pain of illness is always governed by social and institutional learning. For example, pain has different explanations in different societies (Zborowski 1969), and it is reported before one enters into the second stage of illness, such as the sick role (Parson 1951, p. 129).

There is a need to examine the perception of Hadis embedded in their social conditions. Illness is felt and negotiated every day, according to the perceived notions and threat to the physical capacity to attend routine work. I have attempted to study Hadis' perception of illness and health that passes through various stages of illness. The tendency to delay in reporting and consultation to a health care provider and practitioner is also governed by social and financial conditions.

1.2.2 Fieldwork Journey: An Intermittent Travels

I went to the field in Hadi Cooli for fieldwork in 2008, after a gap of 8 years since my previous field trip, to collect some field data for my M.Phil. degree. However, during this period, I kept visiting the community and connected myself with the people and most with the youth and elders. Whenever my field was possible, I could make some time to engage with Hadis and also spend time in their daily routine hours in order to keep my observations alive about the life and lifeworld of the households and workers engaged in manual scavenging, presently sanitation workers. The intensive fieldwork for this work began by re-establishing my contact with them and initially exchanging pleasantries and enquiring about their work, household, etc. It was to continue for a longer period of time, unlike earlier when my travels were to keep myself connected with the community. Subsequently, I met various people at different points in time in Hadi Cooli and wrote brief notes about them. This fieldwork began under the supervision of my research guide, and the communication began between researchers and the supervisors by writing field notes that brought initially path-breaking comments to raise further questions about the field and the Hadis there. During the fieldwork, I was observed as the only outsider visiting the locality of Hadis[5] as I learnt that other caste members do not like to visit Hadi Cooli and maintained a social distance between Hadis and others in Hadi Cooli's vicinity.

The data on the socio-economic conditions of Hadis provided an opportunity to investigate the impact of health services on availability and accessibility and various forms of impediments to access health care services. Understanding these areas helped to understand the perceptions rooted in their routine life and activities that help to relate to illness experiences.

Over a period of 5 years, I conducted unstructured, in-depth interviews with 18 people whom I met on more than 20 occasions. While interviewing and meeting the Hadis, some of them gradually became my key informants. My fieldwork period was prolonged from January 2008 to July 2013 but not in a consistent manner over this period. During this period, I used winter and summer vacations and other personal leaves to conduct the fieldwork. I was unable to stay constantly. However, I could utilise an entire year or more due to my limitations at the workplace. However, I could utilise a maximum of my vacation and leave period in the field for a couple of months at a time, especially during summer vacations and a couple of weeks during the winter holidays.

An in-depth interview was conducted with Hadis, who were available, shared valuable data, and discussed it with me at length. Of 18 respondents, 10 became key informants with whom I would meet and spend hours chatting about various issues. I interviewed members from each household to gather household composition census data on 51 Hadi households. However, the constant source of data collection remains those available even after I returned from fieldwork. Five Hadi elders found my discussions and simply being around them, visiting them frequently, enjoyable. I wrote case studies on key informants and other community members who were interviewed.

[5] During entire fieldwork stays, I was one of the few who went inside the streets of Hadi Cooli, sat on streets and houses whenever and wherever families allowed to converse by sitting at their premises. It would be quite common to stop over and sit onto their open spaces. Interestingly, people would inquire and find out about research purposes, reasons of being there, and what it would bring to them in return. These questions have also been raised and noted by earlier ethnographies and in most studies based on qualitative data.

I also attended two meetings of the Hadi youth. These meetings are exclusive and confidential. They do not allow outsiders to attend the meetings. Santu Hadi introduced me to the Hadi youth during one of the meetings that were in progress in 2010. They were pretty hesitant to see me suddenly inside the hall along with Santu Hadi. Gradually, I introduced myself and told them about my research. When I saw pictures of Subhash Chandra Bose and Mahatma Ramesh in their community hall, which is within the Hadi Cooli, I talked about them with the participants of the meeting during and after the meeting of youth. They were less aware of these personalities. These were donated by a Member of Parliament (MP) when he inaugurated the community hall. My discussion about them helped me to get some space in the exclusive meeting of the youth organisation. We also discussed Dr. B. R. Ambedkar whose picture was not in their community hall. In 2014, when I revisited the community hall, I could see a change; Ambedkar's picture was hung on the wall of the hall. It was perhaps in the context when I inquired why Dr. Ambekdar was not in their hall and whether they were aware of him.

1.2.3 Focussed Group Discussions

I conducted three Focussed Group Discussions (FGDs) on various aspects of Hadi Cooli. As the houses had less space, I sat in familiar places and outside the houses on the narrow streets, where people would join and/or stop by during my conversations with Hadis in and around Hadi Cooli. The first FGD was conducted during the winter of 2009 in front of Ramesh Hadi's house. It was themed on identifying and mapping Hadi Cooli in Chas. The resources available in and outside the Hadi Cooli were identified. Such details exercise was interesting to them as it was the first time an outsider was asking those details. They felt happy to talking about these issues. Hadis identified a well near the government primary school which had dried up. Another hand pump near Noori Masjid in Muslim *Mohalla*, quite close to Hadi Cooli, had also dried up. Identification of temples, common places, places of defecation and latrines, ponds, doctors' clinics and medical shops, and grocery shops in and outside the Cooli was mentioned and pointed out to me. Ganesh Hadi began moderating the discussion and cross-checking certain data with the FGD members if there were doubts. My younger brother, Alam, who had joined the Master's in Social Work programme at the Central University of Rajasthan, was of great help in conducting the FGDs. He also moderated people to open up on the relevant data.

The second FGD focussed upon their occupation, caste and inter-caste relationships, social interaction, changes in the area in recent years, and so on. The third FGD focussed on the health care facilities, availability, and problems in seeking treatment. Bokaro Steel Plant employees, Nagarpalika (corporation) sweepers, MADA[6] employees, and those employed in other sectors also joined the discussion. The meeting covered all types of health services available: private and public. It was also

[6] Mines Areas Development Authority (MADA).

focussed on why they chose some health providers more often than others. Around 8–10 members participated in FGDs. Many of them participated in all the FGDs. It helped me to cross-check the earlier data, if there were some doubts and clarification needed. After returning from the fieldwork, I took the liberty of sending my brother whenever I needed some clarity to clear doubts during my study leave, 2013–2014 A.D.

I interviewed the doctors who have been health care providers at the Hadis. Two doctors who are practitioners in the neighbourhood locality were interviewed. They were interviewed and met on more than a dozen occasions whenever I visited them. One of them practices at Bauri Cooli (a neighbouring Cooli of another Scheduled Caste, known by their name), and became my key informant during and after the fieldwork period. I could call him over the phone whenever I needed any data clarification. His clinic is located at Bauri Cooli. Whenever I was free and had time to spend, I would sit in his clinic. While going to Hadi Cooli from my place of stay (as I was unable to get a room in Hadi Cooli despite my efforts to rent one), I had to pass by his clinic. The main entrance to Hadi Cooli has one grocery shop in Ramesh's house. While walking from the North direction towards the south side to Hadi Cooli, one cannot miss the clinic on the approach road before Hadi Cooli. Therefore, I also got acquainted with a few Bauri residents from Bauri Cooli. I witnessed the interactions between the doctor and his patients at this clinic.

A medical shop where Hadis seek medicines for most of their minor illnesses became my second place for observation after the clinic. I sat there for long hours and observed the visiting patients at the shop. In the initial period of fieldwork in 2008–2009, the medical shop owner was hesitant and would not communicate much about my research. I started meeting him in June 2008 when I planned to visit Hadi Cooli. I spent a short duration, a month's stay, in Chas but barely got acquainted with him. Even later on, after I completed the fieldwork in 2013–2009, we developed a good rapport that continued throughout the entire period of fieldwork. Even later on, after I completed the fieldwork in 2013, in 2011 we realised having a distant kinship link. My mother's sister's daughter was married to his father's brother's son in Bokaro. However, he remained a key informant providing various bits of data that later helped me to develop the chapter on perceiving health care services through *suya-goli*. This chapter derives data from patients visiting the medical shop. Hadis and the medical shop owner's interaction provided various ways of narration illness. Most women would send their children to buy a few tablets rather than visit themselves. In the same chapter, the data collected at the clinic became a vital input for the chapter.

I also interviewed an Ayurvedic doctor in the main market who practices in his own Ayurveda medicine shop. The shop functions as the doctor's clinic as well. His son manages the sale of medicines over the counter while he prescribes them. Hadis do not visit him except in one case. The doctor and his son behaved warmly with me when they learned that I had come from Hyderabad because, at some point in time, they visited one of their relatives who stayed in Hyderabad.

1.2.4 Key Informants

A few key informants were from other Cooli and Mohalla like Bauri Cooli¸ Shakaro Cooli, and Ansari *Mohalla*. Gulab of Shakaro Cooli enjoyed helping me and visiting Hadi Cooli whenever he had no class to attend in the early year of fieldwork. Alam introduced me to Gulrez, who owns a bookstore in Ansari Mohalla, who helped me a lot at the beginning of my fieldwork. He knew a few youths of his age group in Bauri Cooli and Hadi Cooli. Alam had some friends in the main market who had shops in the main market. Fortunately, I too discovered my cousin sister's husband who owned, I, too, discovered my cousin's sister's husband, who owned a shop in Mahavir Chowk market. However, I was not comfortable visiting him quite frequently. I spent time with them whenever I was free, and they were close to Hadi Cool.

Initially, I intended to do fieldwork in Bauri and Hadi Cooli. This changed at an earlier stage of fieldwork play. Some Muslim and Bauri students play and go to the common school. It helped me to connect with a young boy (age 19) who I learned was the son of Doctor Vimal Singh. He introduced me to his friends from Bauri Cooli and a few young boys from Hadi Cooli during one of my initial visits to Hadi Cooli. He later became a resourceful person whenever I needed his presence as he was better conversant in *the Khortha language than myself the Khortha language than I*. Our first visit to Mahuar Basti to meet Hadis on the outskirts of Bokaro was also fruitful. In two subsequent visits, I village. He was always curious about my research and how I would be writing it. During the fieldwork period, he was pursuing his undergraduate graduation and wanted to be admitted either to Delhi University or Jawaharlal Nehru University as he got to know that I am familiar with both universities. Later, I arranged one entrance application form for him in 2011. Therefore, he continued as a key informant of Hadi's youth. I also benefited from the discussion with friends of Suhail and Chas Medical Shop in the later phase of my fieldwork. He would introduce me to his other friends and about my research. It became almost a regular practice to meet new people at his medical shop, and I would explain my research. At times, it would be tedious and repetitive to me.

I tried to take a room on rent, Hadi Cooli, but failed. Most Hadis could not give a room on rent as their houses had limited space and had at most two small rooms and a veranda. There are several houses only one room and a veranda. I had an opportunity to with one room for a month in 2010 under certain conditions. It was near Hadi Cooli in a house on the main market road. Later, I was asked to vacate the room, as the owner stated a shortage of space in his house. I wanted this space to use some time to write field notes and talk to people from the field. Later, one of my key informants told me that Hadis's visit to his house made the owner of the house uncomfortable. However, he could not tell me this reason directly as it seemed inappropriate.

Most of my interviews were during the late afternoon hours with working Hadis. During forenoons, I conducted interviews with women at Hadi homes and a few youths who would be sitting outside the houses and playing cards, ludos, etc. during the winter season. One respondent, also a key informant, who retired from the Steel Plant, would mostly be available at his house. Even when he was unavailable, his son was available at the grocery shop. Quite a good number of Hadis would use the same street where I would be sitting. So, quite often, a few Hadis would stop over and start talking to me. A partially blind woman was very vocal in most of our meetings. She would feel happy whenever I visited her. She is usually not given importance, and people do not often talk to her.

I had several interviews with patients from the Hadi community when they visited the clinic in the neighbourhood. During the interviews, I learned that Hadis make little use of government hospitals and health centres. I raised the non-availability of doctors and free medicines to the poor in my meetings with the Deputy Commissioner (DC) and the district civil surgeon (CS). I was also helped by an old friend from Delhi who was an IAS officer in Jharkhand. He arranged my meetings with the Deputy Commissioner of Bokaro to gather some data on Bokaro. The DC also arranged a meeting with the civil surgeon. Later, it became convenient to meet the CS and also interview doctors at the PHC and sub-divisional hospitals. Subsequently, I met most officers whomever I wanted to interview and collect some data to save time. I never visited the Block Development Office (BDO) at Chas prior to the fieldwork except once during my earlier fieldwork. Samar Foundation (an NGO) helped me travel to remote villages in my initial periods of fieldwork within the Bokaro district.

1.2.5 The Field

In my research, the term 'field area' means the Hadi Cooli, a locality constituted exclusively of Hadi, a Scheduled Caste in Chas town in Bokaro, as mentioned earlier. I chose Hadi Cooli in Chas town my field Vinay Srivastava writes, 'To the place where the members of the community the investigator plans to study and reside' (Srinivas 2003, p. 11). Therefore, Hadi Cooli in totality was the field area in my research. The intensive fieldwork method was followed. It also included the case study method. As the study began, the most important technique worked out was to hold unstructured interviews with the respondents once a rapport was established. This helped significantly but was not the only way to gather data.

The idea that a fieldworker should involve the knowledgeable in the community appeared significant, but the most appropriate way to involve all potential respondents, and they were valuable and informative. My study is significant as an initiative to study Hadi Cooli and follows the anthropological maxim that the fieldworker should try to study a community as a whole (Srinivas et al. 2002). In my case, being a Muslim, entry into the life and history of manual scavengers from Muslim families was quite difficult except that picking up the local dialect was easier for me in the field. As aptly described by Tulsi Patel, the practical aspects of a researcher, and she writes that fieldworkers' skill is tested in the field (Patel 1994). Hadis had inhibitions about my intentions as to what I would be doing after collecting data about them for some study. They never recall anyone like me who had come to do research in this way, being around for a long duration, and having constant interest in all or most activities they would do. Hadis could recall a few people asking briefly about their household details and nothing more. I was also an outsider to them as my parents settled in Bokaro much after the Steel Plant came up but were not linked to the BSL. After an extended period, the Hadis realised I was not asking questions that made them uncomfortable to answer. At times, my presence was not starkly visible, and the fieldwork started giving me data on various dimensions and issues of interest, even without asking a specific question.

The introductory writings in the book by Srinivasan and Mohanty (2008) guided me to conduct intensive fieldwork. The influence of the fieldworker is lesser than the community members who are dominant in the interaction process in certain ways. However, the techniques and knowledge of the fieldworker also prevail in the interaction process. Both these situations arise in the field. I thought to raise questions pertaining to the life of Hadis not only on the basis of my observations and data but also the earlier studies that have discussed, the lower caste (Shah 2002; Thorat and Newman 2010; Acharya 2002, 2010a, b), especially menial occupational castes. Secondary data were also used from Chas Municipal records and the Census of India.

I chose Chas town in Bokaro as my field area because of my familiarity with the poor living conditions, a certain sense of social stigma, menial and traditional occupations and discrimination faced by Hadis. Hadi Cooli (the habitat/settlement where the Hadi caste lives) is a primarily marginalised urban locality, avoided by most social groups cutting across religious lines. These circumstances of the Hadis strengthened my resolve to take up this social group for intensive ethnographic study and come to a deeper understanding of the Hadi lifeworld.

My research project examines the socio-economic status of the Hadis and uses that to understand illness, illness perceptions, and the utilisation of health care services among Hadis. In one of the chapters, I also explore the socio-demographic profile and attempt to sketch the occupational status in light of their traditional occupation and contemporary jobs.

In this above-discussed context, the study had certain objectives to be explored in the intensive fieldwork and examination based on the data collected. The idea to understand the perceptions and notions of illness and health among the Hadis in Chas district Bokaro, also explored the process and structure of accessing health care facilities in an urban setting and contextualise in their traditional occupations and studied the illness experiences and access to health care at various treatment centres and examine the factors influencing their behaviour.

Because of their low socio-economic conditions, their environment, and working conditions, the Hadis have a higher risk of infection and disease. Hadis are residents of the Bokaro district, which was carved out of the erstwhile Dhanbad district. The discussion on Hadis' life and various other aspects is also discussed concerning Dhanbad district since there are a good number of Hadis who live in Dhanbad and have kinship ties with Hadis in Chas town.

Moreover, it is essential to keep the background that the field is part of a new state that was created only in November 2000, constituting 18 districts of the erstwhile state of Bihar that were separated into the state of Jharkhand. Bokaro district is one of 18 districts in Jharkhand. The selected pocket, Chas Town, is one of the

Municipal towns of the Bokaro district. Bokaro district consists of two subdivisions, eight Development Blocks, and 136 Panchayats comprising 639 villages. Chas is one of the subdivisions of Bokaro, having one sub-divisional hospital. Even before the separation of Bokaro district from Dhanbad in 1991, Chas had the civic status, and its civic affairs were under Mines Area Development Authority (MADA). The status of Chas as a Municipal town has also helped in the continuation of the traditional job of Hadis to the contemporary period.

1.3 Overview of Chapters' Arrangement

Chapter 2 describes the study population in Hadi Cooli in Chas town, in the district of Bokaro. This attempts to elaborate on the research problem, its significance, and how it relates to the Hadi population. I also elaborate on the discrimination faced by the Hadis in the town. Their deprivation in each sphere of life speaks volumes of the crude reality in comparison to other caste groups in the locality. I will also be using studies on caste discrimination that discuss aspects of health care services. It is significant to keep in mind that there have not been any previous research studies on the Hadi caste in this area of research, which would make this relevant in a broader understanding of illness and health practices in a regional context. As mentioned earlier, they have been engaged in manual scavenging occupation for years, and individual land records available with a few households show that they have been engaged in manual scavenging occupation for years. Individual land records available with a few households show the 200-year-old history of Hadis in Chas town, Bokaro. Hence, the study also helps to understand and relate the conditions of the Indian untouchable, like Hadis as manual scavengers and formerly untouchable, to learn about the problems and perceptions of those who have been the classic example of social deprivation and segregation due to traditional association with menial and cleaning occupation. The experiences and perceptions that the untouchable 'Dalit' castes like Hadis can provide (Khare 1984) the data on the present social situations and in my research, the oral social history of Hadis is also discussed.

Chapter 3 discusses field data on various aspects of illness and health-seeking behaviour. The vital essence of illness and illness experiences lies in Hadis' lives and their interaction with health care providers and institutions vis-à-vis accessibility to health services during illnesses. The chapter 'Illness Experiences and Perceptions' analyses these experiences and understands the process of construction of perceptions in illness. The illness narratives among Hadis provide various dimensions embedded in the community's traditions and practices.

At the conceptual level, illness and health are two concepts that are embedded in the same cultural phenomenon. The individuals' subjective feelings get translated into some kind of disease or disorder in the human body. The feelings are explained with the help of cultural symbols, symptoms, and words. For instance, the Hadi youth from the Hadi caste express the inability to work when they feel uneasy and unable to maintain their normal bodily capacity. The routine of going to work early in the morning is affected. This may be due to high temperature, severe body ache, not finding rest after work, etc., but these symptoms are not considered feeling ill and worthy of seeking clinical treatment.

In my study, the patients from Hadi Cooli felt that they did not receive human dignity from doctors and administrative staff during visits to Chas Hospital and Primary Health Centres (PHCs). Several respondents felt uncomfortable visiting government hospitals. Hadi patients have little role to play except abiding by doctors' prescriptions in order to get proper investigations (Kleinman 1988).

In such settings, the existence of patients has become least important in terms of patients' choice to go under investigations or delay or refuse and further to be informed about the investigations. Cultural barriers to getting social accessibility to health care services at the centres create a difficult situation for Hadis. Although the minor investigations do not cost a high amount, those tests will be referred outside the hospitals by the doctors. That increases the burden of treatment. Hadi patients need to invest more than 1 day in visiting Chas Hospital and getting the treatment at the cost of losing one or a day's wages. Whereas the private clinics and their local doctors, like the RMPs, do not immediately refer them for investigation, which provides them comfort and does not stop them from going to work. We find that the concept of 'total institution' is not merely a social fact but has been accepted as a larger reality in the modern bio-medical developed regimes. The entire focus of the hospital is to attain common objectives, and the administration needs to behave commonly with all the associates of the institution (Goffman 1961).

Chapter 4 is profoundly engaging on 'Perceiving Health Care Services through *Suya-Goli*'. It is relevant to develop a health-seeking behavioural understanding of health-seeking behaviour and the different stages of illness. Suchman (1965) presented a theory of stages of illness and medical care that discerns five stages demarcating critical transition and decision-making points in medical care and behaviour. The five stages are symptom experience, assumption of the sick role, medical care contact, dependent-patient role, and recovery or rehabilitation.

After a decade, Kosa and Robertson (1969) writes that two comprehensive views gained wide acceptance among medical authorities and laypeople regarding illness and health. One view views illness as any state that has been diagnosed by a competent professional, while the other widely accepted view defines illness in patients' terms: whoever feels ill should be regarded as sick.

Chapter 5 is an important area that provides greater detail about the health care services and how Hadis face impediments while seeking treatment. Understanding the determinants of illness and looking into the impediments of health care services is a significant area in my thesis. Discussion on 'Socio-Economic Impediments to Health Care' in Chas town examines the kinds of impediments Hadis encounter. The health-seeking processes not only demand certain experiences to be discussed but also require analysis in relation to various forms of impediments. Health expenses of Hadis have been an important factor that also deters them from seeking treatment outside Hadi Cooli.

Chapter 6 draws the conclusion of the dissertation. The primary intention to study the perceptions of illness and health among Hadis posed challenges to me. I was unclear on how to develop my arguments and arrive at a conclusion. The division of chapters demonstrates the process of inquiry into the questions of illness and health perceived among my study population. The concluding chapter brings the core area of the dissertation to analyse them briefly. To keep the continuity of the research problem progressing in the dissertation, the next chapter begins with a description and analyses the data on the social history of Hadis and their caste-based traditional occupation.

References

Acharya, Sanghmitra S. 2002. Health Care Utilisation in Rural North India-A Case of Nirpura District Meerut. An unpublished Project Report on '*Monitoring Shifts in Health Sector Policy in South Asia Project*'. New Delhi: CSMCH, JNU.

Acharya, Sanghmitra S. 2010a. Caste and Patterns of Discrimination in Rural Public Health Care Services. In Sukhdeo Thorat and Katherine S. Newman (eds.) *Blocked by Caste-Economic Discrimination and Social Exclusion in Modern India,* pp.208-29. Delhi: Oxford University Press.

Acharya, Sanghmitra S. 2010b. Access to Health Care and Patterns of Discrimination: A Study of Dalit Children in Selected Villages of Gujarat and Rajasthan. *Children, Social Exclusion and Development, Working Paper Series,* 1 (2). New Delhi: Indian Institute of Dalit Studies (IIDS).

Acharya, Sanghmitra S. 2011. Conceptualizing Social Discrimination in Access to Health Services towards Measurement: Illustrating Evidences from Villages in Western India. *Indian Emergency Journal*, VI (II): 14-27.

Acharya, Sanghmitra S., Mala Mukherjee., and Kanhaiya Kumar. 2015. Social Inclusion in Universal Access to Antenatal Care: Understanding the Inter Group Inequalities. *Journal of Social Inclusion*, 2 (4), New Delhi: Indian Institute of Dalit Studies (IIDS).

Aiach, P. and Curtis S. 1990. Social Inequalities in Self Reported Morbidity: Interpretation of Data from Britain and France. *Social Science and Medicine.* 31 (3): 267-74.

Anderson, Ronald M. 1995. Revisiting the Behavioural Model and Access to Medical Care: Does It Matter. *Journal of Health and Social Behaviour.* 36 (1): 1-10.

Banerji, D. 1981a. Measurement of Poverty and Undernutrition. *Economic and Political Weekly.* 16 (39): 1579-1582.

Banerji, D. 1981b. The place of Indigenous and Western Systems of Medicine in the Health Services of India. *Social Science and Medicine.* 15A (2): 109-14.

Banerji, D. 1981c. *Poverty, Class and Health Culture in India.* New Delhi: Prachi Prakashan.

Benyamini, Yale. 2011. Health and Illness Perceptions. In Howard S. Friedman (ed.) *The Oxford Handbook of Health Psychology*. pp. 285-318. New York: Oxford University Press.

Bhardwaj, Ruby. 2010. Medical Pluralism in India: The Interface of Complementary and Alternative Therapies with Allopathy. In Arima Mishra (ed.) *Health, Illness and Medicine: Ethnographic Readings*. New Delhi: Orient Blackswan.

Blane, D. *et al.* 1996. Illness Behaviour and the Measurement of Class Differentials in Morbidity. *Journal of the Royal Statistical Society.* 159 (1): 77-92.

Bowling, A. 1991. Measuring health: A review of quality of life measurement scales. Open University Press.

Bury, Michael. 1982. Chronic Illness as Biological Disruption. *Sociology of Health and Illness.* 4 (2): 167–182.

Carstairs, G. M. 1955. A village in Rajasthan. In M. N. Srinivas (Ed.), *India's villages*, 36–52. Government of West Bengal Press.

Chaplin, Susan E. 1997. Scavengers: Still Marginalised. *The Administrator.* Vol. XLII (January-March): 123–151.

Charmaz, K. 1999. Stories of suffering: Subjective tales and research narratives. *Qualitative Health Research*, 9 (3), 362–382.

DiMatteo, M. Robin and Freidman, Howard S. 1982. A Model Course in Social Psychology and Health. *Health Psychology.* 1 (2): 181-193.

Dimiuou, N. 1995. Illness and Culture: Learning Differences. *Patient Education Counsseling.* 26 (1-3): 153-7.

Djurfeldt, Goran. and Stefan Lindberg. 1980. *Pills Against Poverty: A Study of Introduction of Western Medicine in a Tamil Village.* New Delhi: Macmillan India.

Elliot, J. 2005. *Using Narrative in Social Research: Qualitative and Quantitative Approaches.* London: Sage Publications.

Faircloth, C. A., Boylstein, C., Young, M. E., and Gubrium, J. 2004. Sudden illness and narrative flow in narratives of stroke recovery. *Sociology of Health & Illness*, 26 (2), 242–261.

Frank, Arthur W. 1994. Reclaiming an Orphan Genre: The First-person Narrative of Illness. *Literature and Medicine.* 13 (1): 1-21.

Frank, Arthur W. 1995. *The Wounded Storyteller: Body, Illness, and Ethics.* Chicago: University of Chicago Press.

Frank, Arthur W. 1997. Illness as Moral Occasion: Restoring Agency to Ill People. *Health: An Interdisciplinary Journal of Social Study of Health, Illness and Medicine.* 1 (2): 131-148.

Frank, Arthur W. 1998. Just listening: Narrative and Deep Illness. *Families, Systems and Health.* 16 (3): 197-212.

Funch, Flemming. 1995. *Transformational Dialogues: Facilitator Training Manual.* No. 1. California: Transformational Processing Institute.

Goffman, Erving. 1961. *Asylums: Essays on the Social Situations of Mental Patients and Other Inmates.* Chicago: Aldine Publishing Co.

Gregory, Richard L. 1987. Perception. In R.L Gregory and O L Zangwill (eds.) *The Oxford Companion of the Mind*, pp. 598-601., Oxford: Oxford University Press.

Grover, Shalini. 2007. *Looking Inside Marriage: Lived Experiences, Notions of Love and Kinship Support amongst Working Women in New Delhi.* Working Paper Series No. E/285. New Delhi: Institute of Economic Growth.

Goswami, B. N., and Xavier, P. (2005). Dynamics of "internal" interannual variability of the Indian summer monsoon in a GCM. *Journal of Geophysical Research*, 110. [Note: Volume/issue/pages not available—if you have them, please share for completion.] ResearchGateNatureACP

Hannay, David. 2011. The Symptom Iceberg. *British Journal of General Practice.* 61 (584): 224.

Haron, Yafa., Eisikovits Rivka., and Shai Linn. 2004. Traditional Beliefs Concerning Health and Illness among Members of the Circassian Community in Israel. *Journal of Religion and Health.* 43 (1): 59-72.

Husserl, Edmund. 1970. *Crisis of European Sciences and Transcendental Phenomenology.* Evanston: North Western University.

Ilaiah, K. 1998. Why I am not a Hindu: A Sudra critique of Hindutva philosophy, *culture and political economy.* Samya.

International Institute for Population Sciences (IIPS) and ORC Macro. 1995. National Family Health Survey (NFHS–1), 1992–93: India. *IIPS.*

International Institute for Population Sciences (IIPS) and ORC Macro. 2000. National Family Health Survey (NFHS–2), 1998–99: India. IIPS.

International Institute for Population Sciences (IIPS) and ORC Macro. 2007. National Family Health Survey (NFHS–3), 2005–06: India. IIPS.

Julia E. *et al.* 1989. Health Perceptions of Primary Care Patients and the Influence on Health Care Utilization. *Medical Care.* 27 (3): 99-109.

Karlekar, M. (1976). Poverty and women's work in India. Concept Publishing Company.

Kelly, M., and Field, D. 1996. Medical sociology, chronic illness and the body. *Sociology of Health & Illness*, 18 (2), 241–257.

Kulkarni, P. M., and Baraik, V. K. 2003. *Utilisation of health care services by Scheduled Castes in India* (IIDS Working Paper No. 39). Indian Institute of Dalit Studies.

Kumar, V. (2005). Situating Dalits in Indian Sociology. Sociological Bulletin, 54(3), 514–532. http://www.jstor.org/stable/23620623

Khare, R. S. 1984. *The Untouchable Himself: Ideology and Identity*. Cambridge: Cambridge University Press.

Kleinman, Arthur. 1988. *The Illness Narratives: Suffering, Healing, and the Human Condition*. New York: Basic Books.

Koos, Earl L. 1960. Illness in Regionville. In Dorrian Apple (ed.) *Sociological Studies in Health and Sickness*, Pp. 9–14. New York: McGraw Hill.

Kosa, J. and L. Robertson. 1969. The Social Aspects of Health and Illness. In Kosa et al. (eds.) *Poverty and Health: A Sociological Analysis*, Cambridge: Harvard University Press.

Labov, W., and Waletzky, J. 1967. Narrative analysis: Oral versions of personal experience. In J. Helm (Ed.), *Essays on the verbal and visual arts*, 12–44. University of Washington Press.

Last, J. M. 1963. The iceberg: Completing the clinical picture in general practice. *The Lancet*, 282 (7297), 28–31. https://doi.org/10.1016/S0140-6736(63)92717-6

Massey, James. 1997. *Down Trodden: The Struggle of India's Dalits for Identity, Solidarity and Liberation*. Geneva: WCC Publications.

McAteer, H., Elliott, J., Hannaford, P. C., and Heaney, D. 2011. Evaluating web-based self-management support for people with chronic pain: A randomized controlled trial. Pain, 152 (3), 833–842. https://doi.org/10.1016/j.pain.2010.12.008

McBroom, William H. 1970. Illness, Illness Behavior and Socio-economic Status. *Journal of Health and Social Behavior*. 11 (4): 319-326.

Mechanic, David. 1968. *Medical Sociology: A Selective View*. New York: Free Press.

Mechanic, David and Edmund H. Volkart. 1960. Behaviour and Medical Diagnoses. *Journal of Health and Human Behaviour*. 1 (2): 86-94.

Meleis, A. I. 1990. Theoretical nursing: Development and progress (2nd ed.). J. B. Lippincott.

Mendelson, Cindy. 2002. Health Perceptions of Mexican American Women. *Journal of Transcultural Nursing*. 13 (3): 210-7.

Mishler, Elliot G. 1986. *Research Interviewing: Context and Narratives*. Cambridge: Harvard University Press.

Misra, Babagrahi. 1969. Sitala: The Small Pox Goddess of India. *Asian Folklore Studies*. 28 (2): 133-142.

Morrell, D. C. and C. J. Wale. 1976. Symptoms Perceived and Recorded by Patients. *Journal of the Royal College of General Practitioners*. 26 (167): 398-403.

Mumby, D. K. (Ed.). 1993. *Narrative and social control: Critical perspectives*. SAGE.

Murray, S. A., Kendall, M., Boyd, K., and Sheikh, A. 2005. Illness trajectories and palliative care. *BMJ*, 330 (7498), 1007–1011.

Nicholas, Ralph W. 1981. The Goddess Sitala and Epidemic Smallpox in Bengal. *Journal of Asian Studies*. 41 (1): 21-44.

Pargament, Kenneth I. *et al.* 2004. Religious Coping Methods as Predictors of Psychological, Physical and Spiritual Outcomes among Medically Ill Elderly Patients: A Two Year Longitudinal Study. *Journal of Health Psychology*. 40 (4): 126-134.

Parsons, T. 1951. The social system. Free Press.

Patel, Tulsi. 2008. Stigma Goes Backstage: Reservation in Jobs and Education. *Sociological Bulletin*. 57 (1): 97-114.

Patel, Tulsi. 1994, 2006. *Fertility Behaviour: Population and Society in a Rajasthan Village*. New Delhi: Oxford University Press.

Pelto P. J. and G. H. Pelto. 1997. Studying Knowledge, Culture and Behavior in Applied Medical Anthropology. *Medical Anthropology Quarterly*. 11 (2): 147-63.

Petrie, K. J., L.A. Jago and D. A. Devcich. 2007. The Role of Illness Perceptions in Patients with Medical Conditions. Lippincott Williams and Wilkins: *Current Opinion in Psychiatry*. 20 (2-3): 163-167.

Qadeer, Imrana and Dunu, Roy. 1989. Work, Wealth and Health: Sociology of Workers' Health in India. *Social Scientist.* 17 (5/6): 45-92.

Qadeer, Imrana. 1995. Primary Health Care: A Paradise Lost. *IASSI Quarterly.* 14 (1&2): 1-20.

Radley, Alan., Julianne, Cheek and Christian Ritter. 1997. The Making of Health: A Reflection on the First 10 years in the Life of a Journal. In *Health: An Interdisciplinary Journal for the Social Study of Health, Illness and Medicine,* p.141.London: Thousand Oaks.

Ramaswamy, Geeta. 2005. *India Stinking: Manual Scavenging in Andhra Pradesh and their Work.* Chennai: Navayana.

Sachchidananda. 2002. *People at the Bottom: A Portrait of Scavengers.* New Delhi: Concept Publications.

Sahu, S. K. 1986. Social Dimension of Health of Tribals in India: A Case Study of Oraons of Orissa. In Buddhadeb Chaudhuri (ed.) *Tribal Health: Socio-cultural Dimensions*, New Delhi: Inter-India Publications.

Seale, C., and Charteris-Black, J. 2008. The interaction of age and gender in illness narratives. *Social Science & Medicine*, 62(10), 2577–2590

Shah, A.M. 2002. The Dalit 'Category' and Its Differentiation. *Economic and Political Weekly.* 37 (14): 1317-1318.

Shyamlal. 1984. *The Bhangis in Transition.* New Delhi: Inter India Publications.

Smith, Judith A. R. N. 1981. The Idea of Health: A Philosophical Inquiry. *Advances in Nursing Science.* 3 (April): 43-50.

Srinivas, M. N. 2003. An Obituary on Caste as a System. *Economic and Political Weekly.* 38 (5): 455-459.

Srinivas, M. N., A.M. Shah and E. A. Ramaswamy. 2002. *The Fieldworker and the Field.* New Delhi: Oxford University Press.

Srinivasan, K. and S.K. Mohanty. 2008. National Family Health Survey 3- Household Deprivation and Its Linkages with Reproductive Health Utilization. *Economic and Political Weekly.* 43 (48): 67-76.

Suchman, Edward A. 1965. Stages of Illness and Medical Care. *Journal of Health and Human Behaviour.* 6 (3): 114-28.

Thorat, S., and Sabharwal, N. S. 2006. Caste, social exclusion and health [Working papers/essays on social exclusion; exact title varies across editions—commonly cited IIDS/UNICEF work]

Thorat, S., and Newman, K. S. 2010. *Blocked by caste: Economic discrimination in modern India.* Oxford University Press.

Thurston, E. D. (1909). *Castes and tribes of Southern India* (Vols. 7). Government Press. JSTOR

Wale, N. J. (1976). Sociology and health care. Macmillan.

Webster, A. Merriam. 1993. *Webster's Third New International Dictionary.* Chicago: Encyclopedia of Britannica.

Weiss, G. L.,and Lonnquist, L. E. 1997. The sociology of health, healing, and illness (3rd ed.). Prentice Hall.

World Health Organization (WHO). 1978. *Declaration of Alma-Ata.* WHO/UNICEF International Conference on Primary Health Care, Alma-Ata, USSR, 6–12 September 1978.

Ziyauddin, K.M. 2022. Situating Hadis' Occupation and Caste. CASTE: A Global Journal on Social Exclusion 3 (2), 263-284

Zola, Irving K. 1973. Pathways to the Doctor - from Person to Patient. *Social Science and Medicine.* 7 (9): 677-689.

Zurbrigg, Sheila. 1984. *Rakku's Story: Structures of Ill Health and the Source of Change.* Bangalore: Centre for Social Action.

Chapter 2
Stigma and Survival of Hadis in Chas and the Residues of Erstwhile Manual Scavenging

2.1 Introduction

Located in Hadi Cooli in Chas town, near Steel City, Bokaro, Jharkhand, the Hadis are an occupational group that forms a sub-caste of the Scheduled Caste and has resided in a municipal town for many years. The socio-cultural milieu of the Hadi community places them at the bottom of the caste hierarchy, making them one of the most disadvantaged caste groups among the 22 Scheduled Castes noted in the Census of India documents from 2011. Dhanbad and Bokaro are the two districts with the largest concentration of Hadi families, approximately five hundred households, as described by the president of Hadi Yuva Samaj Manch (HYSM), Ramdhir Hadi. Significantly, the Hadis are not recognised as a distinct caste group in the Scheduled Caste (SC) list of the Census of India document for the state of Jharkhand; instead, they are recorded as a sub-caste of the Mehtar caste, which was historically a dominant manual scavenger community in the region. Hadis have been categorised among the traditional occupational caste groups and are commonly known as Hadis, meaning a community engaged in menial and sewage-cleaning work. So far, the Government of India has provided the 2011 Census of India as the enumeration list. Hadis are grouped with Mehtar and Bhangi in one category, distinguishing themselves as manual scavenging and sweeping caste groups in Jharkhand.

This ethnography on the Hadis of Chas town primarily explores their cultural life-world to inquire about the Hadis' perceptions of illness and health, helping to frame the study within sociological perspectives of health and illness and to contextualise it. The Hadis represent one of the most significant social groups within the broader category of Dalits. Consequently, the study explicitly examines Hadis residing in Chas town in the Bokaro district, excluding other regions and neighbourhoods. Numerous accounts of anger and discrimination are evident, particularly regarding the failure to accurately categorise the Hadi caste. Despite their persistent requests to census officials to classify the Hadis separately, they have repeatedly been grouped with others, as seen in India's last two census reports. Gandhi Hadi lamentably notes that

K. M. Ziyauddin, *Occupational Caste Groups' Illness and Health in India*, People, Cultures and Societies: Exploring and Documenting Diversities,
https://doi.org/10.1007/978-981-96-8967-5_2

"we are put together, but our caste names were also miswritten as Hair rather than Hadi (Census of India, 2001), and further, it turns out to be Hari in a decade (2011), and god knows what would be next!"

For centuries, the studied community has attributed itself to Hadi Jati, rather than as the census has enumerated them. The Hari caste consists of Mehtars and Bhangis combined into one category. An intensive and in-depth data point reveals that most members of Hadi Cooli identify as Hadi and not Hari. Russel and Lal (1916, p. 216) wrote, "The Bengal name Hari is supposed to come from Haddi in Bengal, which is supposed to derive from 'Haddi', a bone, and is the bone gatherer and was familiar to early settlers of Calcutta." They (Vol. IV, p. 217 and Vol. I, p. 367) further wrote "Hadis as the sweepers" and a bone collector or gatherer, a synonym for Mehtar, the sub-caste of Mehtar in West Bengal. Sociologically reviewing, caste names used by families and titles transferred through family lineage have been consciously changed over time. In his study, Srinivas (1942) found that caste names were altered in census records even before British enumeration officials. Therefore, it is unsurprising that Hadis are recorded equally as Hair or Hari in various census documents.

In 2011, Hadis, Mehtar, and Bhangi comprised 58,242 persons in the state. Out of the total population, the urban section of the three castes was reported to be higher at 36,044 people, whereas the rural composition stood at 22,198 in the state of Jharkhand. The field observation data and key informants supported that Hadi, Mehtar, and Bhangi castes are urban populated communities in the state, compared to rural households from these castes. There is an established framework to substantiate how manual scavengers are associated with urban communities because toilet systems evolved in cities and towns, though historically they were associated with the royal houses and the king's palaces in cleaning and sanitation work. Public toilets in modern cities and municipalities expanded the demand for menial work, forcing such caste groups to join this sector quickly.

Together, former manual scavengers from the Bokaro district and modern safai karamcharis account for 10,581 out of 58,242 people in the state. With 5286 in the rural region and 5295 in the urban area, Bokaro boasts nearly equal members from the three aforementioned castes: 5286. At the district level, almost equal numbers of people reside in urban and rural areas. It is important to note that Chas is the most populous town in the Bokaro district. Additionally, Chas town employs Hadi sweepers—manual scavengers. The equal distribution in rural and urban areas indicates that Hadi families from Hadis are settled close to cities in search of employment opportunities and to sustain their livelihoods.

Indeed, small-scale mines and mining operations emerged after the British Empire began exploiting the mineral resources of Jharkhand's tribal land, establishing settlements like Dhanbad, Bokaro, Hazaribagh, Ramgarh, and Ranchi. While these settlements may not meet the definition of a town, a demand for manual scavengers arose when mining companies built housing for their workers and officials, thereby creating a need for labour. The spatial distribution of housing reflected the Hadis population in the Bokaro district, which included both urban (5286) and rural (5295) individuals, with little difference between the two. The scattered Hadis housing in villages, small towns, and district centres reveals a historical factor: the job opportunities available to the Hadis, allowing them to earn a living for their families without the significant burden of migration. This dispersed pattern of Hadis housing provided easier access to employment and proximity to their homes. Initially, this situation seemed beneficial, but later, they faced difficulties in establishing a functional kinship Returning to understand the demographic trend of Hadis in Chas town of Bokaro district, there are 35 wards, with five additional wards added when the previous Chas Municipality was renamed Chas Municipal Corporation on February 9, 2015 (Census of India 2011; Chas Municipal Corporation 2015). Ward number 11 is where the research population resides—Hadi Cooli. With a population of 3632, this ward comprises 646 households. Manual scavenging is illegal to use or promote, as the Parliament Act of 1993 forbade it. The workers refer to themselves as "sweepers" instead of manual scavengers. The most recent Bill, "The Prohibition of Employment as Manual Scavengers and their Rehabilitation Act, 2013" (Lok Sabha Secretariat Parliament Library and Reference, Research, Documentation and Information Service (Larrdis) 2013), offers comprehensive policies and guidelines for rehabilitating such workers. This chapter covers and examines several aspects of Hadis: their settlement, Hadi population, household census, facilities, schools, challenges, work, and daily life, among others, in Hadi Cooli, a small locality in Chas town.

An estimation of the Hadis linguistic population within the overall figures for India reveals that the largest groups of Hadis come from Bengali families, totalling 581,000 Bengali speakers. The secondary languages utilised by the Hadi community include: Oriya 212,00; Hindi 126,000; Rangpuri 44,000; Bhojpuri 38,000; Maithili 4600; Santali 4400; Kui 2400; Kharia 1700; Kurux/Kurukh 1500; and Magahi speakers 1400 (Jhosua Project 2014).

2.2 Hadis Lives Across Borders of Indian States

A community primarily located in the eastern states of India is also dispersed throughout the country in varying proportions. In his study on Bhangis in Rajasthan state, Shyamlal (1981) noted that Bhangis are India's most widely populated and scattered occupational caste groups. In most states, Hadis are categorised as SC, except in Himachal Pradesh, where they fall under Other Backwards Classes (OBC). The Joshua Project reports that the majority of Hadis reside in ten Indian states: West Bengal (589,000), Odisha (224,000), Bihar (106,000), Jharkhand (84,000), Uttar Pradesh (8300), Himachal Pradesh (2100), Andaman and Nicobar Islands (1600), Chhattisgarh (1500), Tripura (1000), and Meghalaya (1000), along with smaller populations both in India and abroad. Additionally, Bangladesh has a Hadis population of 59,000, and their primary language is Rangpuri. A report from the NGO Jhosua Project (2014) notes that they are Hindus. This specific population figure is mentioned because of the Census documents of India, as previously indicated in this chapter.

It is vital to recollect the demographic landscape of the groups engaged in traditional caste-based occupations and find a detailed analysis in this chapter. More to write is more to inquire. Hadis' socio-historical sketch remains largely unresearched and undocumented, unlike many other castes engaged in menial occupations that are widely studied. A comprehensive sociological study of the current community remains an unexplored area of sociological attention compared to all other caste groups nationwide.

Earlier in-depth sociological surveys and studies conducted at various levels and over different years demonstrate curiosity about how Hadis has been overlooked from a significant list of studies, notably a long-term research project by eminent sociologist Shyamlal (1992) on Bhangi of Rajasthan state. An elaborate study on the Bhangi and Mehtar castes titled 'Manual Scavenging in India: A Disgrace to the Country', published under the name of Srivastav (1997), notes a considerable contribution from erstwhile Andhra Pradesh by academic and activist Ramaswamy (2005), who worked for the rights of manual scavengers. She witnessed the demolition drives of dry toilets in the state aimed at eliminating caste atrocities and removing manual scavenging practices. The book was published as a notable work titled 'India Stinking'. Rationally speaking, Caste Hadi as an occupational group has not been reflected much in sociological and anthropological literature in the recent past, except for anthropological compilations like Thurston and Lal (1909). Adding to the irony of Hadis, the people of India series by KS Singh also provides minimal representation concerning Bihar state at that time, with a few mentions of their presence recollected in Census of India has documented data over the last three decades in particular (Census of India 2001, 2011). Occasionally, the government of India has made remarkable contributions to generating a database through various committees and commissions in post-independent India, notably a rich survey initiated by Karnataka state under the chairmanship of IDP Salapa titled 'Improvement of Living and Working Conditions of Sweepers and Scavengers'. This committee report shows that, besides a few Scheduled Castes practising manual scavenging, certain social groups, including Muslims, Christians, Lingayats, Kurubas, and Mudaliars, are also present, making this occupational category a mixed group of workers. Similar to these historical mentions, a recent study on sanitation workers' health, occupation, and lifeworld conducted in Jharkhand and Telangana by Ziyauddin (2022) shows a small presence of non-Scheduled Castes. Some time ago, Venkatesan (2005) reported that states do not acknowledge the persistence of such practices across multiple states (Venkatesan 2005).

Evidence and studies have consistently shown that socio-economic pressures have historically compelled many individuals from castes and communities who were not traditionally manual scavengers by lineage and social classification. This chapter mentions a few youths with tribal identities who have taken up sweeping as their livelihood, highlighting Hadis' presence. Private health care centres, clinics, hospitals, and maternity hospitals/ nursing homes also find tribal youth to be a more valuable and reliable labour force.

2.3 Study Site Selection: Hadi Cooli

Memories of the researcher inform how Hadi community members carried nightsoil as headloads. They would throw the dirt far away in the areas where they worked as manual scavengers, both in practice and manually. Gracefully, the practice of carrying nightsoil as headloads and in buckets has been replaced by the arrival of toilets connected to drains and septic tanks over the last three decades. Social change has been observed following these innovations. Firstly, water-based flush toilets serve as a saviour from the atrocities and inhumane working conditions. Secondly, the formation of Chas Municipality has brought many formal employment opportunities along with social security to the Hadis, which can improve working conditions for selected family members who find regular jobs in the municipality.

An early study on manual scavenging in Bihar by Pathak (1991, p. 37–43) in his book, 'Road to Freedom: A Sociological Study on the Abolition of Scavenging in India', mentions the presence of direct human involvement in cleaning toilets among Muslim countries. It survived for extended periods as Baitul Khula (latrine) was found in Arabian countries, and human excreta had to be disposed of if toilet usage was confined to homes. Pathak emphasises this from a gender perspective, highlighting that Muslim women would use toilets inside their homes and were not allowed to go outside, which led to the persistence of night soil removal practices even in regions of India ruled by Muslims. However, Pathak relies more on his notions of Muslim women practising Purdah and does not support his argument with substantial historical facts and documents. Why do the princely states of other religious communities practice the latrine system in their palaces and royal houses?

In the present-day scenario, flush toilets have emerged as the central toilet system not only in the study's location but in India as a whole; however, the presence of septic tanks remains a reality where waste passes through. The bigger question is: who cleans septic tanks once they are filled and overflow? Until the invention of vacuum-based mechanised machines, the task of cleaning was performed manually by communities who had done this job for generations, cleaning the tanks and dumping the waste elsewhere by hand. This routine of performing such filthy work has evolved into an occasional affair until the septic tanks are filled. Zizek termed this complex situation of menial workers as an extension of cleaning the toilets and continuing occupation (Atluri 2011). The occupational association of the Hadis has changed somewhat over the last two decades due to interventions by non-governmental and civil society organisations that demanded proper rehabilitation with dignity for manual scavengers. The Hadis have also sought dignity in their work and appropriate policy protection for sanitation workers; however, little has improved in their lives.

The researcher conducted an initial field survey of all the localities in Chas town, which helped me identify one Dalit locality that always referred to themselves as Hadis. This decision gradually emerged to select Hadis as a caste-based occupational community for ethnography, and many intermittent visits were conducted over the years. I also realised and reflected that Hadis were one of the most severely affected groups of Dalits (SCs) in Bokaro district regarding their socio-economic standing and nearly nonexistent political participation. Their engagement in the most polluting and menial occupations, such as sweeping and cleaning the filthiest areas of the city's houses and public spaces, significantly contributes to the body of knowledge on Dalits in India. This work's hazardous and filthy nature places it at the bottom of the dignity and labour hierarchy, and no one would choose it as their profession. It also highlights the irony of social distance and stigma in people's lives during social interactions, leading to the notion of physical touch as impure and polluted. Thus, this research emerges as an important study. I selected Hadi Cooli as a location for fieldwork and studying their life ethnographically.

I selected a locality, colloquially referred to as Cooli (Mohalla), as a neighbourhood among all the Coolis inhabited caste-wise in Chas town. I chose to study their perceptions of illness and health through intensive fieldwork. As mentioned, a Cooli in this region is referred to with a specific caste-based settlement belonging to the Scheduled Castes (SCs). Depending on regional variations, this reference can change in different areas within and outside Jharkhand. For instance, the Hadi settlement is referred to as Hadi Cooli, Bauri (caste) Mohalla, while the Bauri Mohalla is known as Bauri Cooli, and Shakaro Mohalla is called Shakaro Cooli. Sociologically, Cooli serves as a marker of caste identity, specifically for those from SCS; it does not appear in the nomenclature of the caste groups' settlements in Chas. Designating Hadi Cooli as a site for fieldwork was a significant decision and a new academic insight, as, apart from a few mentions in the preceding paragraphs, it has not been explored in earlier studies, unlike the Bauris. A long-term stay and qualitative study on the Bauri caste is crucial for observing and understanding another closely related population group living adjacent to Hadis, as noted by Sengupta (1980) and Freeman (2004).

An additional reason to consider the Hadi caste is that even after the establishment of Bokaro Steel Plant as the largest public sector steel unit in India in 1968, accompanied by a significant subsidiary industrial expansion and business-based market growth in the town, the Hadis did not benefit from social mobility or economic prosperity. The chapter highlights the developmental consequences for the lowest strata of long-inhabited communities, with illustrations discussed in various contexts throughout the chapter. It is also important to understand how urbanisation introduced new housing colonies, created diverse employment opportunities and avenues for self-employment, and established new educational institutions providing benefits to Scheduled Castes (SC's). However, it appears that families of the Hadi community remained untouched and did not benefit from these changes, living in their cocoon for reasons that need exploration.

In attempts to understand these factors, the Bokaro district, as a district headquarters and township, undergoes a significant transformation. However, the adjoining municipal town of Chas grows slowly and serves as a subsidiary market for Bokaro Steel City. This outgrowth brings many health care services to the town near Hadi Cooli. The Hadis could witness and experience this developmental process during the entire fieldwork. The book thoroughly explores the socio-cultural world and its group life. The book presents various dimensions of Hadi's life as a community. The following chapters gradually explore their perceptions of illness and health, health care services, and barriers to accessing health care services. My study is based in Chas, a medium-sized town that developed as a service centre near Bokaro and is now contiguous regarding physical growth. Chas is bordered by four rivers: the Govai River to the east, the Garga River to the west, the Damodar River to the north, and the Izri River to the south.

The study population, Hadis, lives on the northern side of the primary market in Chas. On the western side, the river Garga divides the area geographically between Bokaro Steel City and Chas Town. The river serves the large population living along the banks of Chas for bathing and washing. There is a Dhobi Mohalla near the check post where the Garga River to the west supplies water for their daily need of washing clothes. The northern side of Chas borders Dhanbad district and is geographically separated by the Damodar River. Chas is one sub-division of Bokaro district, while Dhanbad is an adjoining district; the southern and eastern parts of Chas border Purulia district of West Bengal. As discussed earlier, Purulia district also has a Hadi population, mainly due to the prevailing customs in the past under the Maharaja of Kashipur. Historically, Chas has been an important town that gradually developed into a regional commercial centre. It is one of the Bokaro Urban Agglomeration (UA) towns, all of which fall under the Bokaro district. Bokaro was established as a district in 1991, carved out from Dhanbad (Chas Master Plan 2027). Chas is strategically situated at the junction of National Highway 23 (NH 23) and National Highway 32 (NH 32).

The study's population is located quite close to the main market area of Chas. It is also important to note the location of Hadi Cooli, as shown in Map 2.1. Hadi Cooli is almost in the backyard of the primary road market. The bypass road of Chas is the busiest route stretching between Mahabir Chowk (the auto stand of Chas) and the check post market, marking the border of Chas and the beginning of Bokaro Steel City (maintained by SAIL). The Bokaro Steel City authority does not allow private and unplanned colonies to develop in the designated area of the plant, which is why most people, including migrants, reside in the neighbouring town of Chas. However, unlike the migrants and others, Hadis have lived there for generations, tracing back to a time much before the establishment of the Steel Plant in the 1970s.

Map 2.1 shows the social mapping of Hadi Cooli in Chas town, prepared after the fieldwork. This attempt highlights the most important landmarks related to Hadi Cooli in the manually prepared map. It illustrates the neighbouring localities to Hadi Cooli and important structures like temples, clinics, and hospitals. The streets and roads connecting Hadi Cooli provide clarity in understanding the location of Hadis in the town, which is quite close to the *Main Road Market.*

It is helpful to portray the population in the following graph (Chart 2.1) and observe its different layers. The chart captures a multi-scalar reality of marginalisation, illustrating how a smaller and deeply embedded caste community is affected and nested within the districts increasingly larger and more centralised administrative units. Structural invisibility is a reality. Small caste communities like the Hadi Cooli families are barely visible in the broader demographic statistics of the town, and caste is embedded not only in rural areas but also spatially in urban spaces like

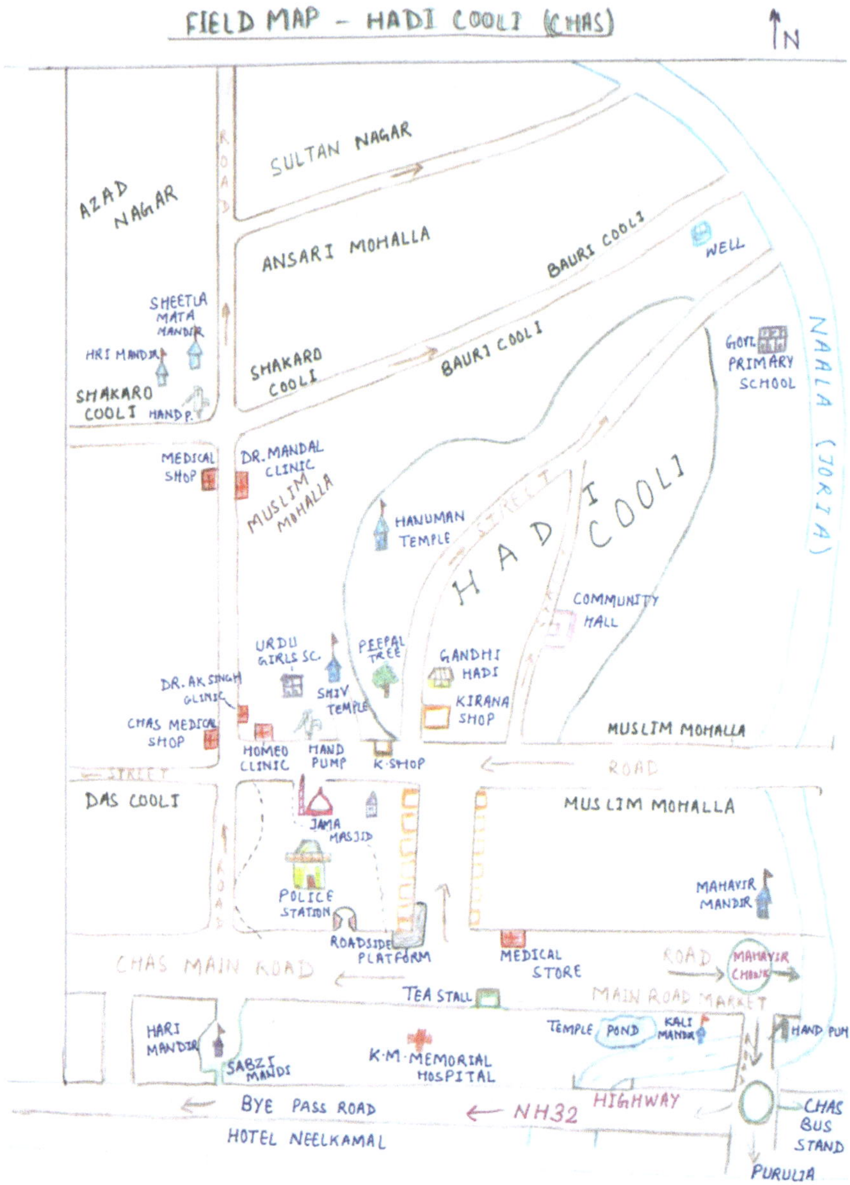

Map 2.1 Hadi Cooli

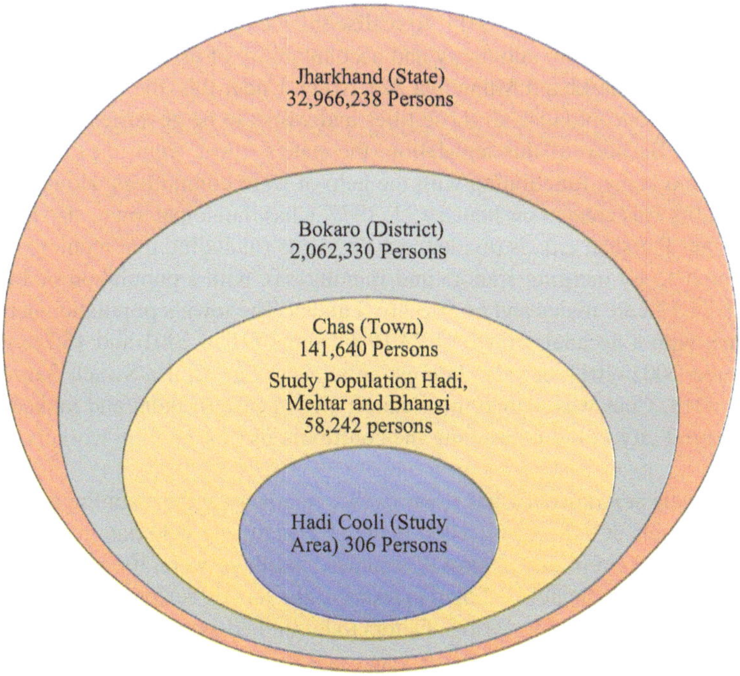

Chart 2.1 Population Hierarchy from Jharkhand State to Hadi Cooli Study Area

Chas. Sociological explorations on illness and health benefit from intersectionality, allowing us to examine how caste intersects with the realities of class, space, and governance. Therefore, the following chart is a quick reflection of the population data.

Jharkhand became a separate entity on November 14, 2000. Additionally, Bokaro was established as a separate district, comprising the Bermo sub-division of Giridih and the Chas sub-division from Dhanbad in 1991. The state has 24 districts, five sub-divisions, and 264 blocks. Chas is one of those five sub-divisions and is a municipal corporation town in the Bokaro district. It became one of the districts with a high SC population in the new state of Jharkhand more than 20 years ago. The district has two sub-divisions, eight development blocks, 136 panchayats, and 639 villages. Chas and Bermo are the two sub-divisions in the district and function as the second administration level. Nawadih, Bermo, Gomia, Petarwar, Jaridih, Kasmar, Chandankiyari, and Chas are the eight blocks in the district. This outlines the levels of governance in the form of sub-divisions, development blocks, and districts.

The All India Town Directory of Census enumeration, 2011, classifies Bokaro town as Class I and Chas as Class II. However, the town of Bokaro Steel City is an Urban Agglomeration. Chas town is a Nagar Parishad (Municipal Corporation). The usage of Nagar Parishad emerged after the state government passed the Jharkhand

Municipal (Nagapalika) Act, 2011, to realise the idea and the "principles of participation, decentralisation, autonomy and accountability of urban self-government at various levels" (Jharkhand Municipal Act 2011). Under this, the new state decided to define and conceptualise all those titles and works to be governed in urban self-governance. Because of this legislation, the state's towns went ahead with urban elections and began functioning with the help of ward committees. However, Chas Municipality was created on January 21, 1977. Chas municipal town, also known as Chas Nagar Parishad city, is divided into 35 wards (including five municipal wards added in 2015 by merging Kandra and Bandhgora), with a population of 141,640. In total, 74,727 are males and 66,913 are females. The town's population density is 338/km^2, with a decadal growth of 23.19% from 1991 to 2001 and 45.7% growth during the 2001–2011 decade. Interestingly, according to the Swach Sarvekshan held in 2018, Chas was named the cleanest city in eastern India and ranked as the 19th cleanest city in India, bringing the contributions of sanitation workers into the limelight.

The female sex ratio of Chas town is 895, relatively lower than the average sex ratio of the state at 948 against 1000 males. By analysing field data of Hadi Cooli, 167 females were reported against 139 males, and there were 306 persons in Hadi Cooli. A positive indicator is shown in terms of the female sex proposition of Hadis, unlike Chas town, bringing various factors to be studied as its attribute. The number of females in the age group of 0–9 years is 43, compared to 37 male members. The figure of the female and male population changes between the age groups of 50–59, as it is 8 for females and 5 for males. In the Hadi community, demographic data indicates a notable gender disparity among individuals aged 60 and above, with 12 females and only six males recorded in this age group. This imbalance reflects a numerical difference and underscores underlying socio-behavioural factors affecting male longevity. The reduced life expectancy among Hadis men is often attributed to occupational hazards and a prevalent culture of alcohol consumption. Specifically, the habitual intake of low-quality, locally produced liquor has been linked to deteriorating health outcomes, leading to increased mortality rates among males. Consequently, the observed higher female-to-male ratio in older age brackets may not necessarily indicate a cultural preference for female offspring but rather highlights the elevated incidence of premature male mortality within the community. This phenomenon aligns with broader patterns observed in rural Indian populations, where studies have documented higher rates of alcohol consumption among elderly men compared to women, contributing to adverse health effects and reduced male longevity (Goswami et al. 2005).

2.4 Health Care Spectrum

The Chas district hospital is known as *Sadar* Hospital. It is located in Sector-1 of Bokaro Steel City (BSC). It is almost 6 km away from Hadi Cooli. Bokaro was made a separate district in 1991; all district headquarters' offices were in Chas town.

By the end of 2000, the headquarters had gradually shifted to BSC Sector-1. Even the residential quarters of officers were located in the BSC Sector 4. Though newly established, the Bokaro district got all the attention and developmental focus. On the other hand, Chas did not get much attention regarding infrastructure development. Therefore, even the possibility of upgrading the old infrastructure was lost during such a shift. Perhaps it was more appropriate for the district administration to shift to the steel city area, a well-planned location, than to construct a new infrastructure in Chas.

Meanwhile, Sadar Hospital could not be developed as a referral hospital in the district. It merely remained as a Civil Surgeon's office and centre of administration to implement health policies at the district level. Chas Hospital, already a Sub-Divisional hospital in Chas Sub Division, got little attention. There has been no improvement in the 10 years after the headquarters shifted to Bokaro Steel City. It is located at a distance of around 5 km from Hadi Cooli. The district administration needed help developing health infrastructure in a town that was proliferating in population and size. Except for this hospital, no other hospital is more prominent in size and establishment in the district. However, it remained a sub-divisional hospital catering to the needs of the rural population as well as patients of Chas town. The number of patients gradually increased, but facilities have stayed the same. The increased burdens of the hospital have affected the health care delivery to patients who travel from faraway locations. One Primary Health Centre (PHC) is located at the Check Post area in Chas, 3 km from Hadi Cooli. Hadis has two referral points for visits in case of serious health issues: the PHC and the Sub-Divisional Hospital on the outskirts of Chas. The location of hospitals is perhaps one of the reasons that deters Hadis from visiting and availing the services of a government hospital. Sadar Hospital in Steel City does cater to immunisation and other health-related schemes of the Government of India. It does not have the infrastructure and is smaller in capacity than the Chas Sub Divisional Hospital. The district Civil Surgeon has an office in Sadar Hospital, and most district officers also function from there. Since I returned from fieldwork in 2012, the district-level planning and implementation of various government policies are managed and controlled by the civil surgeon's office in Sector-1, B. S. City, but it lacks referral services. There is no facility for outpatient department (OPD), inpatient admission (IPA), or most treatments. Therefore, public health centres have restricted access to Hadis to avail free of cost services. This is primarily due to a lack of public transport to and from Hadi Cooli to these places. Also, Hadis cannot make it on foot to these facilities, especially when they are ill.

The distant location of the hospitals has provided space for private practitioners in Hadi Cooli, and they have made their way to most parts of the town. In the vicinity of Hadi Cooli, there are two RMPS, two medical/chemist shops, one homoeopathic doctor's clinic, and one physician (with an MBBS degree). However, Hadis seek most treatment at the RMPs' clinics and the Chas medical shop (over the counter). They also visit Neelam Nursing Home, located at a distance of around 3 km. Generally, during pregnancy and birth-related health care needs, Hadis visit Neelam Nursing Home. Sometimes, a few other private practitioners' services are sought in quite a good number across the town; distance-wise, they are accessible to Hadis. The fees to be paid in private hospitals are pretty high, which also deters Hadis from availing of their services and leaves them little choice but to return to RMPs as their favourite doctors. A few big hospitals located at a distance of 1–3 km are K M Memorial Hospital and Research Centre (KMMHRC) and Muskan Hospital and Research Centre (MHRC). Providing health care to these two hospitals is beyond Hadis's financial capacity.

While locating the institutions of health care providers at a distant location, one finds BGH functioning quite well for the steel plant employees. It is 10 km from Hadi Cooli in Sector 9 of B. S. City. Mostly employed Hadis take advantage of this hospital. A sub-centre also functions at Cheera Chas before the urban governance election, at a distance of around 3 km on the Northern side of Hadi Cooli. It caters to the village Cheera Chas (in Khortha, Cheera means old), which has now become part of the urban ward in Chas Municipal Corporation.

Hadis also takes help from traditional healing practices, such as Dua-Tabeez and Jhad-Phook, and helps a Baba who occasionally visits Hadis Cooli. They also go to other healers like *Ojha*, as discussed in Chap. 3, when Chanda was treated for her prolonged cough and breathlessness outside the Chas town whenever needed in illness or prolonged illness. Therefore, the health care spectrum in the vicinity of Hadi Cooli in Chas town shows the presence of various health providers. The extent to which these are available to Hadis is discussed through their illness experiences.

Integrated Child Development Services (ICDS) near Bauri Cooli is a key centre for children from Bauri and Hadi Cooli. Most children know about the food to be served in the centre and enjoy being present daily for the food. One Urdu-medium secondary school is located in Muslim *Mohalla*, primarily serving students from Muslim families. Ram Rudra High School is located at *Jodhadih More*, which is located at a distance of around 3 km on the eastern side of Hadi Cooli. A few children go to higher schooling at Ram Rudra. The borewell of the primary school near Bauri and Muslim *Mohalla* is dried up. The hand pump on the western side of Hadi Cooli is also dried up. The mosque committee installed this hand pump, but even this one dried up several years back. Therefore, Hadis depend on a tap near Mahabir Chowk and Chas Bus Stand that fetches the needed drinking water for Hadis. However, there are hotels and other people, like truck drivers, who halt in Chas and utilise the water for their purposes.

Chas Bus Stand is an important junction where buses going towards Ranchi, Ramgarh, Hazaribagh, Dhanbad, and Purulia would cross over the stand. The auto stand is located at Mahavir Chowk, just half a kilometre from Hadi Cooli. Many private clinics and practitioners are located in this area as it is a key centre. In the past 10 years, Cooperative Colony in BS City has grown as the most famous private practitioner with their clinic, and a few Nursing Homes have come up there. It is almost 6 km from Hadi Cooli. One does exceptional work with patients from Hadi Cooli visiting the Cooperative Colony.

As portrayed in the social map of the field site, Hadi Cooli was the smallest settlement of a few households in the vicinity of Chas Main Market. It remained in the same housing and spatial pattern until Bokaro was carved out as a separate district, which pushed the developmental activities of Chas town.

Due to the gradual expansion and outgrowth of the city on every side of the town, Hadi Cooli finds itself very close to the central market and residential colonies of the municipal township as a core area. Instead, the bus stand, the police station, the wholesale grocery market, the vegetable mandi (market), and all other public centres appeared to be close to Hadi Cooli, at a five-minute walk. A few private entities or commercial companies started operating. Electro Steel Private Limited and Oil and Natural Gas Corporation (ONGC) have established their units in the city's outskirts, bringing specific promises of developmental activities. A couple of gated housing societies evolved to the tune of metropolitan cities. However, housing in gated communities has not been boosted among city dwellers, who still rely on independent houses and farms. Thereby, the Independent Villa project also did not get its acceptance among the residents and did not grow in the city.

Given the above background, Hadis' families have also found increased employment opportunities in sweeping and cleaning in the private sector, mainly in housing localities, societies, and commercial establishments. However, they have not attended any work at Cheera Chas due to the poor commute and transportation facilities. To them, walking on foot to attend any job would be difficult when the work is available in Nagarapalika and other sectors like private offices, banks, hotels, and hospitals within the town.

2.5 History of Hadis in Chas Region

It was learnt about the king of Kashipur, presently part of the adjoining Purulia district in West Bengal, and he remained a hero in the narratives of the people, as he fought against the mighty British. The struggle against oppression and remaining in a free state was the last effort the Maharaja families lived for. Maharaj Bhuvanaeswari Singh Deo ruled the region until 1947, who governed the Chas region and consequently Hadi workers informed how they lived and liked working under the king's rule than being simply sanitation workers (manual scavengers).

The oral narratives of the elder-most Hadi members mention that the earliest king brought a few members of Hadi Cooli's ancestors and made them settle in the present Bokaro region. From other states of India, and they are not from Jharkhand areas. After getting these people, the Raja (king) provided a particular portion of land to construct houses and perform domestic work for the caretaker of the forts, royal houses, and stables. Land records of Ramesh Hadi's family date back to 1802 AD, which supports their ancestral history in the Bokaro region. Hadis have worked for the Maharaj of Kashipur for centuries and served as workers until the kingdom ended with the declaration of India as an independent country in 1947. The change in the geographical boundary and division of the territory of Manbhum to the Dhanbad region grossly affected Hadi's life and the course of community life.

The Manbhum district became independent in 1833, carving out regions from the Jungle Mahals district, and Manbazar was made the headquarters. In 1938, just 5 years after Manbazar lost its headquarters status, Purulia was made the new headquarters, which was located close to Chas at just 50 km. Manbhum was a larger district in British India that again got divided between Bihar and West Bengal through the States Reorganisation Act and Bihar & West Bengal (Transfer of Territories) Act, 1956 (Purulia District Profile 2014). Hadis remains living and working in the same region, though there have been changes in the nomenclature used and the areas changing from one district to another. Later, Dhanbad was also made a separate district that included the present-day Chas town, changing the domicile of the Hadis from Bengalis to Biharis. Lastly, they got Jharkhandi status after forming the new state of Jharkhand in 2000 AD. Moreover, Bokaro too had already been granted independent district status in 1991. Hadis are a speaking population influenced by Bengali and Oriya languages (Bokaro District Profile 2014). They are found to be settled in Jharkhand as a Khortha-speaking population. Khortha is one of the recognised languages in the state of Jharkhand. Perhaps for this reason, except for one Hadi in Purulia, who later settled after getting a job, all kinship ties are traced within the neighbouring districts of Bokaro in Jharkhand state, but not in Purulia; however, it is not far away in terms of notable distance.

Due to historical facts and a common lineage ruler in the past, Hadis in Bokaro district had common bonds with the families of their caste in Purulia district, West Bengal. However, the historical ties gradually faded between Bihar and West Bengal once they separated as independent states, compounded by the fact that both regions have different linguistic associations, Purulia as a Bengali-speaking region and Bokaro as a Hindi-speaking region. Except for one marriage, no other kinship ties are traced in Purulia. Raju Hadi (Ranjit) is married to Babli Devi, who belongs to a small town, Adra, in the district of Purulia. Gradually, the Hadis of Chas identified themselves as Maghaya Hadi, different from the Hadis living in Purulia, and mostly, they were Bengali Hadis.

2.6 The Land Remembers: Genealogy and Identity in Hadi Cooli

A sociological tracing of the lineage and history of the Hadis settlement helps to understand how this community got into a traditional occupation based on their caste, and they continued doing manual scavenging as their traditional occupation, generation after generation. However, few oral narratives discard that manual scavenging was their traditional occupation; instead, their ancestors worked as caretakers in the royal families of the erstwhile kings of Purulia. Sixty-nine-year-old Ramesh and his neighbour Somesh, 62-year-old Hadi, informed that the kings who ruled the Chas region were basically from the Brahmin (Brahman) caste groups, and Hadis paid due respect to them for their generosity and patronage to Hadis in the entire region. They both said, "*Hamni ke budha-purna ke Raja zameen dale halau aar vehen zameenva par hamnee baes galiye. Baade mein konho soche naa partil ke hamni kae zameen kam aare aadmi boed jeetae. Ab kahan jeebin, badi samasya hau.*

Chas mein konhon zameen nae hau ke hamni liye parbeen. Joria theen kutcho nae hau bachal." A few centuries ago, the local King allowed Hadis to settle on the land provided by the royals, and their ancestors continued their lifeworld in this region. Hadis had no idea this land would become their homeland, and even the land donated to them would appear to be in short supply for future generations. Land provided to Hadis was next to a small natural water stream that has turned out to be a drain, Chas, known as Joria. It is beyond reach for Hadis to buy a piece of land today as the entire surrounding area of Hadi Cooli has evolved into a developed township and a congested marketplace, and residential settlements of migrants have grown over the last two to three decades. We recall how we lived a better, more respectful life during the kings' rule and working for a royal house brought higher respect than other similar caste groups. It is how Shyamlal found in his study on Bhangis in western Rajasthan when Bhangis claimed a higher social ranking among his caste groups who worked for higher than those who worked at the Dhobis castes (Shyamlal 1981).

Ironically, not a single piece of land is available because they can only buy small portions, known as "Gair Majurwa Zameen" (Lahiri-Dutt et al. 2012). In other words, such land does not belong to Hadis or the government revenue department. According to senior elders, these kinds of lands can only be used by government departments for the community's needs and purposes. However, there are occasions when a few people occupy such land until government officials take over for the common good. In this scenario, Hadi Cooli could not expand beyond a very limited land zone due to the unavailability of Gair Majurwa Zameen in the vicinity of Hadi Cooli. Hadi Cooli has been highly congested and confined for years. While it was manageable for the early settlers, the land is currently insufficient for Hadis, forcing them to live in small houses with more family members in one household.

A few Hadis provided detailed data on how revenue collections were conducted by officials of the Raja (king), who were referred to as Misree, 'a colloquially used word for Brahmin using the surname Mishra'. The ruler would send revenue collectors who beat drums while visiting the Coolis and Chas Nagar. Though the Raja had given tax exemptions to the Hadis, unlike the regular tax collections carried out to cover the expenses of the royal families. Apart from running the administration, and with the help of local zamindars, some public services would be funded through the tax amounts collected from the people. Furthermore, the story tells of the bravery of Maharaj, who sacrificed his life fighting against the British invasions in the region. During the British Empire's rule in the Kolkata province, Maharaj was blamed for looting the Purulia treasury, and in due course, he was hanged by the British Empire in Alipur Jail in Calcutta in 19,349 (Purulia District Profile 2014).

In all the narratives provided through oral history, it becomes easier to testify that Hadis worked on the estates of the Maharaja of Kashipur and also for the local landlords (Zamindars) in the Chas Purulia region. Oral stories are passed down from one generation to another. Ramesh Hadi, a relatively active and visibly informed community member, provided detailed data on how gradually Hadis began losing their social ranking and status after the decline of the local Maharaja's empire during British rule and after independence. His forefathers have been telling how the lives of Hadis became stigmatised once they began doing menial jobs in the houses of people in Chas, ultimately facing very stringent and discriminatory treatment in the everyday lives of the workers.

The discrimination meted out to Hadi has been part of the narratives of family members, as a child's identity is rooted in Hadi Cooli, synonymous with manual scavengers. Children were known by their parents' occupation and not by their age and sex among their peers. Ramesh says, "I was not allowed to have a cup of tea at the tea stall in Chas bus stand. If I insisted, the tea would be served in an earthen cup (kulhad). After putting tea in an earthen cup, a hotel helper would be asked to keep it on the floor, and I would pick it up to sip it. I felt degraded and would mostly avoid going to the tea stall."

These facts about Hadis's life appeared most during the discussion during the fieldwork about their past and present. What happened 50 years ago? Has it changed over the years to the present? The narration in the above paragraph briefs the kind of behaviour meted out to Ramesh Hadi during his early years of life. One finds a noticeable change in the hotel and tea stalls, market, etc., since Hadis can be seen sipping a cup of tea at any of the tea stalls in the market. Even the researcher had tea with different Hadi Cooli members, including Ramesh, during the fieldwork. Pintu once offered me a cup of tea in 2010 in our initial interaction at his house in Hadi Cooli. Later, we went to have another cup of tea and visited a tea stall at Mahabir Chowk market, Chas. He does not find such discrimination now as he experienced a few decades ago. Hadis took snacks and food in the hotels in the nearby market and Bauri Cooli, but they did not narrate any differential treatment towards them. At food joints in the market area, there is barely any discrimination towards them. Pintu says, "Now things have changed. What I could see in my early life was that children have a better day to live."

The field data presents similarities between Hadis's past life and present living conditions. There are a few commonalities from the past to the present in a Hadis' life. No doubt, working conditions and the nature of manual involvement have changed over the last few years, unlike those of their forefathers. Shailender Hadi, said, "*Hamni badi ganda raho haliye hau, kaam ganda aare kapadvo nae adheek rahvo halae. Kaam badi halau aur logan bhi kam halthun. Ab hamni ke geedar-butroo badi baes raho hathin badhiya kapda, baes khana aar ki dekhbin kate badal gael hau.*" "We used to live filthy lives at an early age due to dirty work and had fewer dresses. We were also fewer in number than the available workers. Now our children and the new generation live decent and good lives. They can wear washed clothes and eat better food than we; what else can you see as a good change than this?"

There has been no change in work, nature, or socio-cultural dimensions of occupational engagements, nor have stigma and exclusions emerged. Indeed, a form of discrimination against Hadis and the stigma faced in everyday life has changed its face and extent. To an extent, the intensity has reduced. People have told and also verified in the field data that two decades earlier, the toilet system existed as dry latrines and a few would have Sandas (an open dug area in a closed-door compound or room) that always needed daily cleaning to rid away from human excreta and keep hygiene in the houses. Almost every house with such a latrine system had an informal dependency on Hadis, who would clean it every day, and for each household, five rupees were paid. Such a social arrangement of work and renumeration existed in Chas town for a long time until the latrine system was converted into septic tanks. Consequently, everyday cleaning of toilets and filths began losing its demand at the household level in all settlements in the town, making Hadis more distressed during the transition. One finds in the revisits of the fieldwork that almost 40 persons out of seventy-five (75) Hadi male workers had regular employment but through contract agencies under Chas Nagarpalika (municipality) as sweepers. In the decade of 2010, Hadis were called sweepers and kept doing it without any hitch or inhibitions; instead, most of them found it a great idea to work in the Nagarpalika.

Hadis's occupational condition has changed. Those Hadis employed in the Municipality have been provided with corporation-provided *thela* (rikshaw) and other tools to lift and clean the garbage from streets and roads. However, others must still manage with their broom or keep them at their respective offices. A few of them are also provided at private offices where they work. Kamal, son of Ganesh Hadi, said, "*Office jaye time mein jhadu leke nae jeeebin hau. Wahein officva mein raakh de hiye. Kakhno to kutchu tips bhi mil Jae hau jakar se bhadva nikal JA hai, ane to atae door samasya hau jael mein. Yehe karan hau ke hum khush hiyo wahan ke kaam se.*" "While going to my office, I cannot carry my broom, wiper, etc. I keep them at my workplace, Reliance Hotel. I am also paid some tips by the guests who help me to meet my travel expenses. Otherwise, it is tough to attend work at a distance of 5 km daily. I am happier doing sweeping work in a hotel than at Nagarpalika."

In this narration, it is noted that Kamal travels relatively far to his workplace, unlike workers at Nagarpalika, who work within the locality of Chas. He is happy working in a hotel, which gives him a sense of belonging to a better workplace. He gets a chance to talk to the guests arriving at the hotel. He also finds that if he behaves nicely with the guest and cleans the washroom when someone checks into the room, there is scope for getting some tips. Therefore, working at a hotel gives him a chance to work in a good ambience rather than cleaning the filthy streets and roads of different localities in Chas. Kamal Hadi finds a change of place when he returns to Hadi Cooli in the late afternoon. Reliance Hotel falls in the Bokaro Steel City area, a much cleaner and spacious township far better than the areas of Chas. After working for a few years, Kamal does not want to return to the Nagarpalika job. On the map of Chas, one can see Bokaro Airport in the northwest direction. Reliance is away in the direction of the road crossing the airport from Kunwar Singh Colony of Chas. The map also shows a boundary dividing the Chas and Bokaro Steel City area by the Garga River. The map of Chas helps to locate the town's geographical boundary, which is divided into different wards. As shown on the map, Hadi Cooli is next to Noori Masjid and is highlighted with a moon and star symbol.

As previously discussed in this chapter, Hadi Cooli is a settlement established during the rule of a local ruler, referred to as Maharaj, who brought Hadis from neighbouring states over 200 years ago. The land records indicate that Hadi settled in Chas on the donated land of the local ruler during the British Empire, and this land lineage record has been maintained by the family of Ramesh Hadi, dating back to 1802 AD. This group of families shares kinship ties with others in a limited way; however, the majority have common roots, which keeps them united as a small kin group in Hadi Coolu for over a century now. The following chart aids in understanding this tracing of lineage through land records.

In the above diagram, Chart 2.2, prepared during the fieldwork, there is a detailed discussion and cross-verification with the senior-most Hadis, including the Khatiyan holders, Ramesh Hadi. The land records from Ramesh helped prepare and understand that Narayan Hadi was the first person to be given land and settled in Chas in 1802 A.D. As previously mentioned, Khatiyaan is a land record of ownership with lineage to the landholders, and it shows the history of the Hadi caste in Chas. The land papers seen in the field helped us understand that three of the nine members mentioned on the map through land records stayed back in Chas, while the others migrated to nearby areas of Dhanbad and Purulia. The people who survived and settled in Chas are Gadaghar Hadi (community members call him Manu Hadi), Manik Hadi (people call him Makun Hadi), and Ganesh Hadi. The other six members migrated to different localities and regions within the state, as Ramesh Hadi told us. Thirty-six out of the 51 households belong to the three persons mentioned above. In contrast, the remaining households consist of those who came from other localities and settled at Hadi Cooli and grooms who moved in uxorilocally after marrying a woman who is a daughter in the Hadi Cooli of Chas town and settled here.

The two maps of Chas provided here demonstrate differential growth and changes over the last decade. Development activities are also coming through initiatives of the government's plan to push the implementation of the Chas Master Plan Map 2.2, which is expected to be completed by 2027.

The location of Chas, as mentioned earlier, is on national highways NH-23 and NH-32, which connect Dhanbad and Jamshedpur, Hazaribagh, Ranchi, and Puruia in West Bengal. The stretch of national highways in the town area has played a crucial role in the growth of motor garages, shops selling automobile spare parts, repairing shops, and hundreds of subsidiary employments. The trucks carrying raw coal and iron ore from coal mines into the steel plant have to stop outside the Bokaro Steel City, and Chas is the town where they halt until the check post is opened. The finished products and scrap materials are transported outside the plant to cities through the same trucks by road. The entry and exit of trucks in Bokaro are essential to employment growth in the above-mentioned repairing-related sectors. Consequently, many automobile repair shops have come up, but Hadis could not enter mechanical and repair work, unlike Muslims, a community neighbouring Hadis.

Notably, the Hadis in Chas have felt the vast change for two decades, which has grossly affected them, as told by a few Hadis. Somesh Hadi, who had a longer lifespan, describes how Hadis has undergone varied socio-cultural changes in the past several decades. He recalls the patronage of the local king that provided them with a sense of belonging to the area in which they are living now. He adds, "Our elders have done her-baud (huge) and hard tasks while working during the King's rule. No records were kept about our work, yet everyone would get paddy or crop to feed their households. We had fewer dependents. An important difference was in the working conditions, where the jobs were assigned in the house and the agricultural field of Raja." After their elders settled at Hadi Cooli around 200 years ago, Hadis had to work at the houses in localities and settlements of urban Chas.

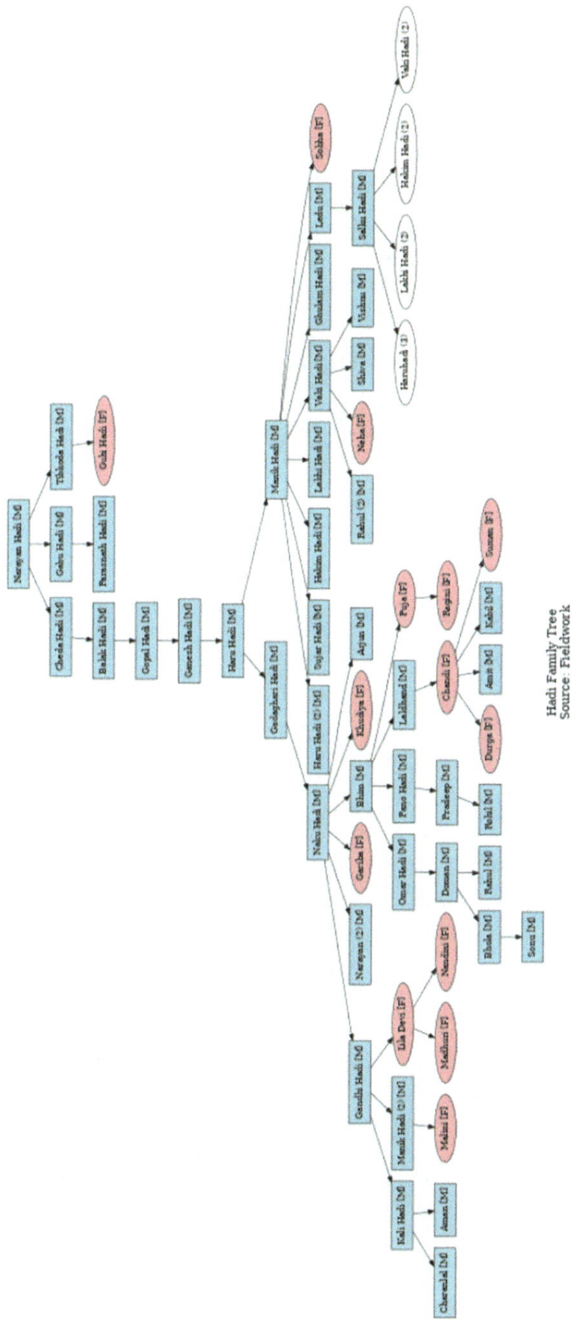

Chart 2.2 Lineage of hadi cooli

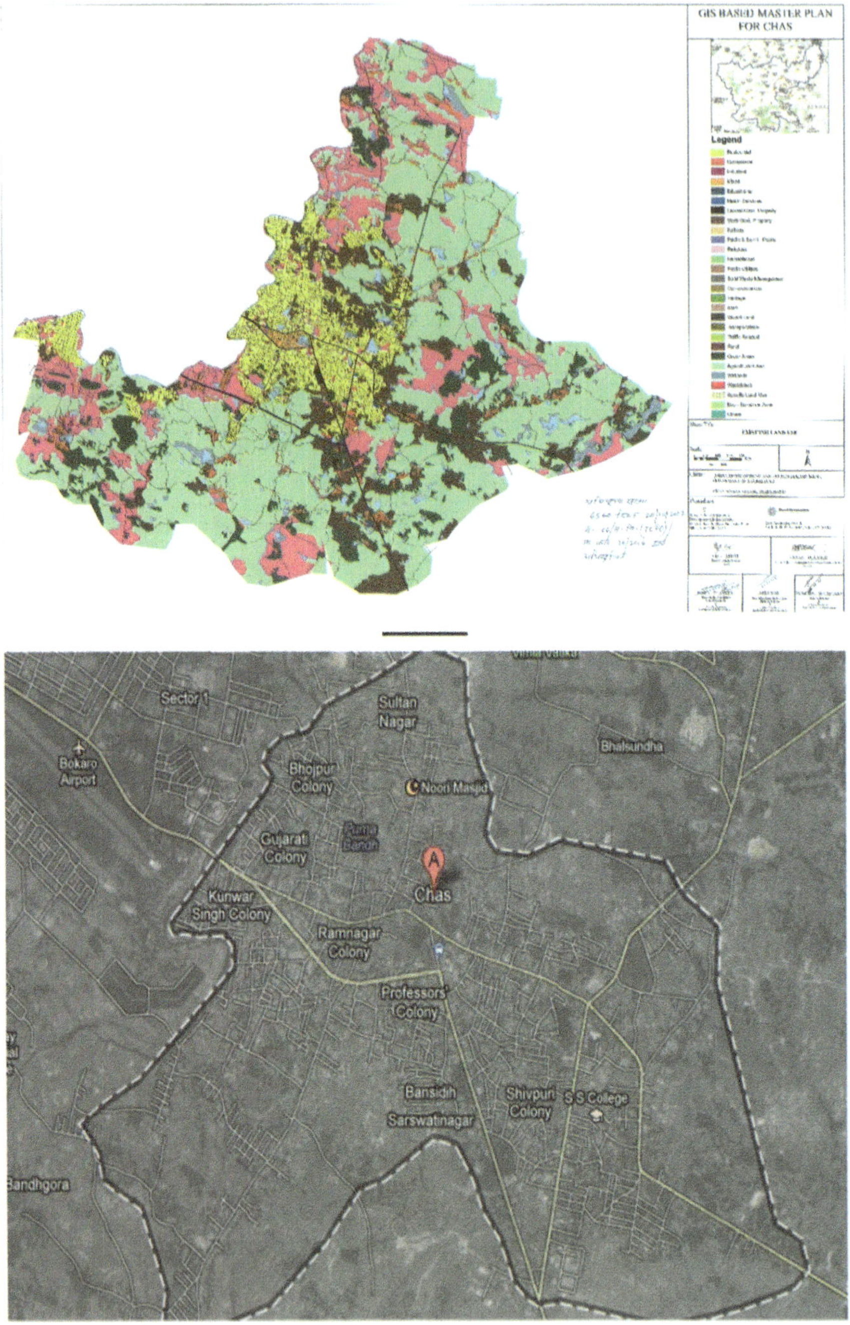

Map 2.2 Chas. Chas MP maps combined.pdf. Accessed 30 April 2025. Source: www.chasmunicipalcouncil.com. Accessed 24 January 2015

Somesh narrated that they would go to work in each *Mohalla* and colony to collect garbage and waste and clean the public drains and toilets of the households. This job was practised around the 1990s or even a few years later. However, the period when dry toilets existed was the worst for Hadis. "We had to carry nightsoil and collect garbage on our heads or shoulders to dispose of it far away in the outskirts of the localities and *Mohalla.* If you believe, I shall tell you that our minds would burst when recalling such inhuman works. Our status has degraded beyond reconciliation and recovery for these jobs and works. There is no way to recover our social status, unlike during the Rajas period, which gave us a better occupational name, *Ghod Sahej*. Otherwise Bauri caste is also Harijan, but then why do people visit their houses and cross the by-lanes into Bauri Cooli, while people do not turn into the by-lanes of Hadi Cooli? Now, you find clean and proper drains in the street. For the past two years, the streets of Hadi Cooli have been cemented, and drains have been sanctioned under the Member of Legislative Assembly (MLA) fund. Otherwise, you could not think of sitting in our street, but we used to eat and cook in the same conditions and live our lives."

He further narrates the work and life of Hadis and helps to relate to the more significant perspectives. He said, "When we compare the income of Hadis with the workers employed as sweepers in the BGH, Hadis stand far lower. For this reason, I tried to send my son to go for higher education. I tell him you may earn less, but your status will change. When I think about work like sweeping and cleaning latrines, I feel so filthy and degraded. Now I can compare because I have seen workers at the hospital doing the same work differently in a uniform and under clean and hygienic conditions. But the working conditions in Nagarpalika are hazardous and low pay." Hadis, in the past, were an alcoholic. Gradually, the liquor consumption is decreasing. Somesh Hadi explained that one could observe many of them drink occasionally. He justified drinking and said, "Without drinking (consuming liquor), Hadis could not perform the job and live in that filth and dirt." Hadis have not entirely abandoned drinking; they still visit Bauri Cooli to consume local liquor at lower prices than a liquor shop in the market.

Now, Hadi youths have somewhat improved their habits and have also reduced their daily alcohol consumption. However, we live under challenging conditions. After Somesh was employed at BSL, everyday problems in the house could be managed as he earns 14,000 rupees monthly. Even today, he can access health facilities and get medicines from Bokaro General Hospital. Now that they have better food, wheat flour (a luxury in Hadi Cooli once) is available to prepare rotis. Those children who eat rotis feel proud and superior and share this happily with their peers. Maybe 10 years ago, they used to keep poultry at home, but nowadays, there is no space left to keep them, and if they keep them, they are hunted by cats and dogs. The number of households and people has also increased, so the space left has been occupied.

The narrative of Somesh helps in observing the changing pattern of Hadis and liquor consumption. He sees a gradual change among many youths, unlike his generation, who were heavily dependent on locally made liquor. This trend of small change is also attributed to the changing nature of traditional occupations. The shift after the Municipality took over the work of sanitation changed the work done every day to a monthly basis. Hadis became a sweeper to clean the common areas under the urban wards, but household jobs remained private work. A household has to pay for services hired for cleaning, etc. The importance of Hadis remains crucial, and it is felt among the other community members, too.

2.7 Hadi Sub-Castes

The studied community, Hadis, is one of the sub-castes among the Mehtar /Bhangi castes on the census list. Hadi as a caste group is divided into four sub-groups, namely;

1. Maghaya Hadi.
2. Sahir Hadi,
3. Digar Hadi and.
4. Bengali Hadi.

Maghaya Hadi are dominat group settled in Hadi Cooli in Chas town. The other three sub-castes who have settled in nearby states have almost no kinship ties with them. These three sub-groups are not part of the present study and do not provide detailed data about their micro life situations. The three other sub-castes appear to have settled themselves in different districts of Jharkhand and have found traces of families in close states like West Bengal and Orissa (presently Odisha). This book does not focus on three sub-castes that have not been studied during the ethnographic explorations, and this demands a focussed study to inquire about exclusive knowledge into their lifeworld. Maghaya Hadis, about the association with local kings, has always asserted and associated themselves with higher social rankings and sought higher social status in marriage and other social ceremonies.

It is reported in the data that mostly Maghaya Hadi kin groups are settled and residing in certain neighbouring districts like Dhanbad, Hazaribagh, Ranchi, Purulia, Giridih, and Ramgarh, and they are close to Bokaro district. It is beyond the capacity of Ramesh Hadi to provide community details by yearly classification and dates of their migration and settlement in Hadi Colony. Social history fact sheets show significant data on how Hadis lived and progressed in Chas town. Gradually, younger children and new generations find it difficult to understand their history and have no clue about their ancestors, unlike the data provided by the senior-most Hadis in this chapter.

"A sad part of the current generation is their ignorance and unawareness that their forefather's social history is our reality, Gandhi Hadi once told and takes a deep breath." I remember barely more than five or six people know about the history of Hadis settlement in Chas, and how few of them provided detailed lineage traces in the initial and later intermittent field visits, that Maghaya Hadi carries out menial work in Chas. They all have a single family root. I visited several other localities, probably to trace other Hadi families in Bokaro, including a more secluded village within the Bokaro township area, Mahuar Bast. Hadi's family had merely some guesses about the ancestors' history, though the elder ones could corroborate the information provided by Hadi Cooli.

In the discussion with the elder Hadis, the group concluded that occupation and work are two different components. Hadis, in a later period, realised that once the British took over the local kings' rule, they were left with few choices for livelihood and household life. The work opportunities were reduced for some time. It was almost unanimously informed data while talking to many senior elders in Hadi Cooli, including Ramesh, Somesh, Pintu, Goli, Haribabu, Jagdish Hadi, Shailender Hadi, and Panbabu Hadi. After the rule of the Raja ended, our elders had fewer choices. Over time, the Raja's relatives left Kashipur, and the Hadi community began taking up other available menial jobs, primarily cleaning and removing excreta. The presence of MADA in Dhanbad created some job opportunities in nearby Chas. In government offices, Hadis continued working as cleaners and were also tasked with disposing of dead animals and carcasses, especially since there was no Chamar caste in Chas urban area. As Goli remarked, "Except for us, most castes owned agricultural land. In such a situation, we remained in menial and stigmatised occupations, passing them down from generation to generation."

In response to the researcher's question, "Why did you take up manual scavenging as your occupation?" Pintu responsded and said, "*Hamni ke itihaas ke jaen ke kee karbeen hao. Ab hamar baap-dada yehe karo halae aare hamro ekre mein laga dal hathin. Ye batvaa hamni kii janbeen. Puche ke Konho reeti naae hau, bus karte jau yehe hau hamni ke badh purna ke niyam.*" "What would you know about our history? My father and grandfather were doing this occupation, putting me into the same. We never thought of the reasons. There is no way of questioning why we chose to practice manual scavenging. Just keep doing our job is what we learned from our elder community members."

The comment answers whether present-day Hadis continued this occupation, and they never thought beyond it. Engaging in scavenging occupations gradually reduced Hadis's social status in the larger social spectrum. It made them an object of pollution in religious notions of pollution and purity, as Dumont studied among Hindus of India. Unlike other professions, the inherited cleaning job does not bring pride and status, unlike many other caste groups. Hadis believed they were a worker in the royal houses and reduced to sanitation workers from being a Ghoda Sahej (caretaker of horse stables) in the past.

However, the above data on occupational changes is given only by elders in Hadi Cooli, whereas the youth are unaware of their occupational background. Maghaya could be equal to a dominant land-owning caste, the Bhumihars in this region, as a few Hadis claimed. Hadis in Hadi Cooli belongs to the Ghoda Gotra. In his 70s, Ramesh Hadi says there were only three houses in Hadi Cooli, three paternal brothers, as discussed later in this chapter. However, an evidential social fact surfaced: there has been a gradual, unintentional, and subtle change in the use of caste surnames such as Hari from Hadi. A few members were found to be using Hari as their surname, while many continue to use Hadi. In the documents of the census of India, they misspelt their original caste surname 'Hadi', and Hadis is mentioned as 'Hair' in the census report 2001, a term unrecognised by Hadis in written or any other reference.

2.8 Occupational Structure: Hadi Working Population

My occupational understanding of scavenging was entirely distinct from my earliest inquiry when I began. I taught it through websites that informed me it referred to a method to remove burnt and exhaust gases from any combustion chambers. Surprisingly, it was not what I was looking for. I then started delving into how manual scavenging is a purely socio-cultural and political affair, making it a functional aspect of Indian society. My study was intended to understand how manual scavenging is a dominant social phenomenon in the entire Indian subcontinent. It is a birth-based inherited occupation enforced through caste hierarchy in social arrangements. Sachchidananda (2002) in his work in Mehtar of Bihar pointed out correctly how the scavenging caste communities never received attention from society, neither from academia in terms of their inception or their existence, desirability, or epistemology.

Vasant Moon, a dalit activist and author, compiled 18 volumes on Babasaheb Ambedkar's contributions in the form of 'Writings and Speeches of Babasaheb Ambedkar' (WSBSBRA), which was published in the year 1989. Ambedkar writes Once an untouchable, he remains always an untouchable, Similarly, Hadis has remained a sweeper once they are born as a sweeper; this expresses the existing contemporary problems and challenges, showing the reality of the Indian deprived Dalit community, who are largely menial workers, like workers and families of Hadis Cooli in Chas. Babsaheb's writing on Dalits currently looks like an old age caste-based occupation persists even in a most fascinating economy. At the same time, workers like sweepers remain merely menial workers, looked down upon in every aspect of their lifeworld. Their status is ascribed, and the birth identity of a caste cannot be changed that way; likewise, all the workers of Hadi Cooli would elaborate in the field.

In addition to this, another aspect is observed. Hadis are not able to change their social status. If they could attempt to change their traditional occupation to newer jobs, there could have been a possibility. Nevertheless, one significant fact emerged from the community. They would have a reverse understanding that curtails Hadis's social and even upward mobility. A firm knowledge among Hadis that their occupation would always be available as a secured employment sector, as none of the caste communities would take up menial jobs like sweeping and manual scavenging in Chas. It is a different matter of fact that in the last one decade, a few a few private hospitals have started hiring youth from tribes and they do not create problem and any threat unlike how Hadis could show their monopoly in sweeping occupation and do not act the dictates of private health sector organziations as per their need. A few hospitals are ready to pick up the tribal youth as they accept low wages and are more loyal workers. The urban local governance has paved the way for better opportunities for manual scavengers in Chas town. Hadis are employed as sweepers in almost the entire employment sector in Chas. The entry of tribal youth is unnoticed by Hadis at present. Perhaps these workers are presently in a tiny number. This data is not described in any discussion in the Hadi Cooli, but the change in the preference to employ tribal youth is found in a few hospitals. The apparent reasons cited are that they are more loyal, stay overnight or at the hospital premises, are less paid than Hadis, and are more regular at work, et cetera, as told by a few hospitals like K. M. Memorial Hospital, Chas.

In all the factual data analysis, prominent discrimination manifested due to occupational engagement in sweeping. Consequently, the practice of social distancing as a measure is a caste-based function of society in Chas. Hadis are no longer required to inform who they are; instead, the word Hadi explains to everyone familiar with the city that Hadis are synonymous with sweepers and a polluting caste in menial jobs. However, it is essential to see how Pandey (2013) discusses the idea of caste and prejudice relevant to the present study in the context of existing prejudice and perceptions towards Hadis by others. In his book titled 'A History of Prejudice: Race, Caste and Difference in India and the United States', a subaltern historian, Pandey (2013), writes how the caste system would have survived and shows the continual existence of prejudice against the oppressed in a latent and manifest way. He deploys the terms 'vernacular' and 'universal' to locate the notion of prejudice, where the former is local, localisable, relatively visible, and even acknowledged. For instance, prejudice against untouchables, indigenous populations, and other minorities is a case in point in the forms of blacks, gays, Jews, and women. In the context of Indian society, his writings stress that prejudice often tends to stem from deep undercurrents of casteism and Hindu social order.

Harshe (2014) illustrates another form of prejudice mentioned by Pandey (2013) that is more or less 'universal' and referred to as 'natural'. It appears unseen but derives familiar aspects of contemporary culture, state, and its laws. Such prejudice is not often recognised. In reality, the power and privilege that perpetuate the practice of caste-related prejudices are based on the existing underlying structures of power and privilege.

A few Hadis do recall how severe forms of discrimination existed, but over time, many forms of that discrimination have declined. Somesh, in his 60 s, while sharing about his childhood, said, "*Hamni konho naae man karo hai yehe kama kareke. Kintu hamar baap aar hamni jakhne jawan haliye toe aare ganda kaam karao halthin. Moodiya ya Kandhva par tatti uthaye ke wanae Joria mein feko haliye. Ghara se ya officva se. ab to kam se kam eee nakhar kaam nah hau. Mana bheen-bheena jaa hovat halae par yehe sach halau. Hamar zindagi tanik baes bhalae, jabae plant mein naukri lagle hamar. Choaed kee tonhon yaad karvon heen. Kintu hamni kutchu bhae jeebe hai to Hadi jaatee ke nae. Doosar jaatee ke ankhiyan mein neeeche raho hiye aar naae izaat do hathin.*" "We never felt like doing this work. However, when I was young, I started working along with my father in the dirtiest job. We had to carry human excreta in a bucket on our heads or shoulders. I would feel like vomiting and have a nauseating feeling when I would carry it as a head load. But gradually, I accepted the reality. My life and working conditions improved after I got employed at Bokaro Steel Plant. Leave it now. Why recall the dirtiest phase of my life? But whatever we change into, ultimately, I would remain as Hadi only. In the eyes of another caste, we are always lower and degraded, and other castes do not give us any respect."

This data reveals what Gyanendra Pandey (2013) argued in his book 'A History of Prejudice: Race, Caste, and Difference in India and the United States'. He says the structure of prejudices that have been built over centuries is quite challenging to disappear in a short time. He says the emancipatory struggles of the lower castes (Dalits) have a long way to go before they can uplift substantial portions of Indian society from practising vicious prejudice. It survived not only based on caste structure but also various other means that perpetuated caste inequality and structure in Indian culture. The existence of manual scavengers shows how the caste structure has survived.

In my fieldwork, Hadis were manual scavengers, now called sweepers, but they still belong to menial and polluted castes, who do a menial job as their traditional occupation. In the past, several castes attempted to change their caste from one to another. M. N. Srinivas, in his published paper on 'An obituary to caste system', illustrates other changes brought about by technological changes and how those products are transforming the lives of villagers governed by caste and traditional occupation. He described the changes brought by new technologies in occupations like barbers, shifting into market and cash-based work, and washermen opening laundries. However, a few castes could not adopt new technological changes due to limitations. For instance, Hadis has not been able to transform their work because sweeping jobs are to be done at the workplace office and home, and attending to jobs involving garbage and removing the same from streets and roads. The change can be brought gradually, such as working, using gloves and tools, cleaning the trash and drains, etc. Sweepers could never follow the better way of some castes. The nature of Hadis's work demands that they visit the workplace, unlike a barber or a washerman. He describes how those products transform the lives of villagers governed by caste and traditional occupations. A few examples are, "Edible oil is now produced in factories, and they have rendered the oil presser and his bullock-drawn wooden press. The barber and the washerman, two essential castes, have had to change their working styles" (Srinivas 2003, p. 457).

This change occurs in that caste where occupations supporting technological inventions have emerged, and technology is proving a way to reduce manual job handling. However, less focus and effort have been put into menial jobs, so finding a technological way out is essential. However, a lesser quantum of change is noticed in the work of manual scavengers (earlier), Hadis' sweeping job. The continuation of sweeping occupation has perpetuated discrimination even in the vicinity of one of the most advanced cities of modern India. Where the first major steel township was established, a certain amount of discrimination has remained in the adjoining town of Bokaro, Chas. However, one does find changes taking place. It is also true that community members who seek help from sweepers would call Hadis, and ultimately Hadis is the Mehtar for the local settlements of the town. Calling Hadis only because they are sweepers and not by their names manifests a particular kind of prejudice against a special group of workers as a community in Chas, and they are Hadis.

The ethnography of the community helps to understand how unequal and hierarchical treatment of workers from the Hadi community continues, sustained by a few youths in their 20s. Notice that the forms and crudity of discrimination against Hadis have changed throughout one generation's experiences. Two prominent members of Hadi Cooli informed how they have also noticed direct discrination against Hadis have reduced due to emerged new market, food and restaurants and roadside eateries are in large numbers where it is not always possible for those units to identify and prohibit Hadis to enter into premises or serving in different plates unlike in their childhood things were quite inhuman. The visible presence of Hadi youth in the market areas after they finish their work shows a kind of space created with higher urban growth and a new market. In a few temples, Hadis are also allowed to enter, though there are two specific temples meant for Hadis, and they are the central community members visiting these two temples.

The new tool to change their surnames and keep Kumar in children's names could blur the future social identity, not Hadi. A few have already started using Hari as their caste surname, which does not provide the social identity of a particular caste group; however, it remains to be seen how social processes take place over time in the town. This social fact demands further inquiry to examine and understand whether it is a deliberate attempt or a normal social act. Unlike the study of Tulsi Patel's paper, 'Stigma Goes Backstage: Reservation in Jobs and Education', which demonstrates in Delhi that there is a concise set of initiatives and ways among the first-generation parents who had worked as sweepers (*safaiwale*) in private and government offices. They used a technique to hide their caste identity as they did not want to go through similar humiliations to their children, which they had experienced and gone through humiliations.

Tulsi Patel writes, "It is not surprising that people from the SC category react in ways to avoid their traditional caste or surnames and use the relative caste neutral surnames, such as Kumar, Pal, Ram, Lai, Chand and Nath. They also prefer to use surnames used by higher castes to hide their caste identity and the anticipated ensuing humiliation in life. Some of these surnames are Singh, Vyas, Guru and Charan. Parents, especially fathers who have been in jobs through the reserved quota, have given their children their first names as surnames to conceal their quota category identity, such as Kapoor, Chand and Swaroop" (Patel 2008, p. 104).

Substantial data exists on Hadis's attempts to change their surnames to hide their caste identity. Their existence is similar to their caste name, which identifies their occupation. A relevant social fact emerges that all forms and natures of ill-treatments and unpleasant experiences at the workplace are directed at the lower castes in India. Data does support the fact that an untouchable would always remain an untouchable and the lowest in the caste rankings in India. Forms of unpleasant experiences at the workplace would surely differ from one person to another. Caste rankings are a reality even today and derive from the stratification system that functions quite systematically (Michael 2007).

As a caste group, Hadis could not benefit from the state's available welfare schemes, likely due to their unawareness and hesitation in approaching the relevant offices of the development block in the Chas division. Hadis always perceive themselves as belonging to the lowest and inferior caste, which results in officials not responding to their requests. Hadis manifest and identify themselves when approaching government offices. It is also true that in the past, similar caste have been referred to by different nomenclatures such as 'untouchables' and 'Harijans' (a glorified term coined by Narsimha Mehta that Mahatma Ramesh popularised), 'Exterior Caste' (a concept used by J. H. Hutton), and 'Depressed Classes' (another term used in British officials' records).

Nayak (1995) wrote in a paper that one member of the Brahmin caste would never do jobs like cobbler, manual scavenging, or collecting animal carcasses. Similarly, an untouchable member would not be allowed to pick up the Veda in their hands and recite. Hadis were never allowed to perform prayers and recite religious text in temples until the fieldwork was completed in Chas, though they are allowed to enter the premises these days. The notion of pollution and purity is provided differently in his write-up based on caste ranking and social distance facts. Aiyappan, in one of his research studies, writes that the Nayadis in 1937 (see Srinivas 1952); the caste members of the Nayar caste of Kerala would stay at a distance of 7 feet if a Nambudri Brahmin were there before them. Further, one Iravan person should stand at a distance of thirty-two (32) feet; one Cheruman caste person would stay at 64 feet, and a Nayadi between 74 and 124 feet, as the other castes (Nayak 1995). It is pretty weird to witness if it exists in the contemporary world and is unconstitutional in India. Much to the credit of the inclusive efforts of the independent Indian system of affirmative policies, most of these practices are not reported elsewhere or anywhere.

Such strict social distancing is diluted and is not visible in this crudeness anywhere among the Hadis of Chas town today. In all group discussions and interviews, I could find different forms of discrimination due to the occupational engagements that turn upon other aspects of Hadis's life, but never a principle like that mentioned in Nayak's study of Kerala. Goli Hadi, 52 years old, works as a sweeper, narrated, "During my childhood, in all eating joints (Dhabas), I was not given space inside the hotel. Tea was not served directly by hand by the worker at the stall. Most tea stalls would keep a *kulhad* (earthen cup). Tea was served in the same cup to everyone, but our cup would be kept at some point, and I would pick it up. My son does not believe me when I tell him about my childhood stories." Similarly, Ramesh Hadi, 69 years old, recalls the discrimination at all the public places outside the Hadi Cooli during his youth and childhood days. Unlike today, he could not imagine entering the house of any other caste group; he at least had access to SCs houses for purposes other than attending to jobs. People do not bother much about castes, unlike in his early life when going to a hospital, bank, or hotel.

A few women also work in the nearby localities, adding some income to their households. The major employers are Sahus or Baniyas and Bengali households, but Hadi women only get sweeping and cleaning jobs in upper-caste Hindu and non-Hindu households. A few women have worked in Sikh households, but the women do not prefer it. The Sikhs' residential area is four to 5 km away, making it difficult for Hadi women to reach on foot.

A census of the Hadis household shows that 113 Hadis members participated in fieldwork during the age group of 0–14 years, and this is the total community

Table 2.1 Distribution of Hadi workers by sex and age

Sex Age group (years)	Male	Female	Total
15–24	18	10	28
25–34	21	9	30
35–44	23	3	26
45–54	5	5	10
55 and above	8	3	11
Total	75	30	105

Source: Household Census Data of Hadi Cooli

members. This number includes both sexes of the community members, males and females. Sixty-one (61) youth in the age bracket of 15–24 years are reported, whereas eighteen (18) males and ten (10) females are available as working members. Further, among the 53 Hadis in the age group of 25–34 years, almost half, 21, are working men, and another nine are reported as working women. The number of working women is decreasing in the age group of 35–44 years, whereas only three women are shown as working compared to 23 men, among the total 34 Hadis in this category. As shown in Table 2.1, the workers' distribution aged 15 years and above shows no under 14 years as working members.

Significant data emerges from Table 2.1 about Hadis's negligible social and occupational mobility. Except for one member who does not carry out menial and sweeping jobs among the 105 workers, all are involved in their caste-based occupations at different offices and sectors. Jhomda Hadi, a 33-year-old person, stood out as an odd man among his peers. He took a private office job in a firm called Tata Indicom during the field collection days as a clerk in the office. It was a telecom service providing outlet in the fanciest market area of Bokaro Steel City.

Jhomda talks that he would spend from morning to evening hours in his office, unlike all his peer group and friends from Hadi Cooli, who return home in late afternoon after finishing the daily cleaning job of sweepers. It is a great matter of pride to him that he could find one comfortable office and a decent job that does not talk about his caste identity. The customer does not ask for his family background, except for the services needed at the office. His office is equipped with good technical support and air conditioning, which he never imagined would be available at his home or any of his community members' houses. He asked a question once in an interview in the fieldwork, "Why does birth in a certain family and caste restrict us from the freedom of choice in all aspects of life and work?" Most Hadis, as workers, are overly dependent on available jobs and are drawn into this menial occupation.

Although it gives both a sense of security and over-dependency, the result is rarely helpful in bringing about a change in their life. Jhomda asserts, "I am the only Hadi who has completed graduation (B.A.), and I am not doing my caste-based occupation, unlike my friends and peers in Hadi Cooli." This is true of caste-based communities that are engaged in the same or similar occupation for generations, which is much more prevalent among the low castes in the region, including the Hadis. The Hadis are not exploring other avenues of jobs or other jobs as such. Instead, they find it easier to get into sweeping jobs. It is the sector where there is little chance of competition from other castes than Hadis. The social milieu in which Hadis have retained their traditional occupation and have been unable to find a way for social mobility. His life in Hadi Cooli is entirely different from his time at work.

Hadis' existence is reflected in the real-life situations and a form of continuity in manual scavenging work, which are perennially performed by menial workers forming my study's population. Hadis fall into a sweeping category, but this nomenclature does not fit into manual scavenging. Their employment is primarily in government offices as Safai Karamcharis, and this journey began after the formation of the Municipal Corporation in 1977. Hadis also had great hopes and aspirations when they found a separate state promising jobs and employment to all natives, but things did not progress well. Ultimately, Hadis have to earn their livelihood in urban offices, private and public sector banks, private hotels and lodges, and guest houses. Hadi Cooli members informed that some Hadis Cooli members found a limited number of jobs in the early days in a few central and state government institutions in Dhanbad town, apart from the already formed Mineral Area Development Authority (MADA).

MADA started functioning as a Municipal Corporation to fulfil the needs of urban and coal regions in Dhanbad district, including Chas, until the Chas Municipality was established on January 21, 1977, and changed into Chas Municipal Corporation on February 9, 2015. In Chas, the development is unplanned and organic. It is focussed along the three main roads of the town, with considerable mixed development along these roads, with housing, hotels, offices, etc., coming up along the roads. There is heavy congestion and even crowding along the roads, especially along the road to the North (Chas Master Plan 2027).

2.9 Hadi Cooli Households

According to the composition of the households' census data collected in 2010, there were 51 households in Hadi Cooli. Hadis' total members comprise 306, which belong to only 51 houses, whereas 139 belong to males and 167 are females, as shown in Table 2.2. The sad story of the data emerges about the number of widows in the community.

A whopping number of 19 widows are reported in such a small community population, and only one woman is divorced. Dhulia, as a widow, is 65 years old and resides with her son and daughter-in-law. Dhulia is younger sister to Ramesh Hadi. Another woman, Bijla Devi, a single woman, lives with her two unmarried daughters. Her husband, Kamal, deserted her and left to settle in Ranchi city. He married a woman and stayed with his second wife at her father's house. No widower was found at the time of data collection. This data also illustrates that men's average life expectancy is shorter than women's in Hadi Cooli, as provided in Table 2.2.

Four of the six surviving older males worked for the Bokaro Steel Plant (a PSU) and have retired, while the other two are employed in private jobs. The first person,

Table 2.2 Age distribution of the hadi population in Hadi Cooli

Age structure	Sex		Total
Age	Male	Female	
0–9	37	43	80
10–19	32	48	80
20–29	23	24	47
30–39	26	26	52
40–49	10	06	16
50–59	05	08	13
60 and above	06	12	18
Total	139	167	306

Lakhicharan, works at ICICI Bank, and Haribabu works at another private office in Chas town. An improved office environment and working conditions, with minimal exposure to filth and hazardous materials at the Bokaro Steel Plant (BSL), as well as in the General Hospital (BGH) of Bokaro Steel Town and the main Administrative Block (ADM) building, have minimised workers' exposure to hazardous working conditions and risky chemicals encountered when entering manholes. Consequently, they report not being alcoholics, unlike most Hadis. Better employment opportunities, regular salaries, fewer health hazards, and free health care services at Bokaro Hospital have benefited some of the mentioned household members and employees, keeping them healthier than their counterparts in Hadi Cooli. A few community elders in Hadi Cooli informed me that the settlement of 51 households stands out as the largest Hadi-populated Cooli region, even in Bokaro's neighbouring districts.

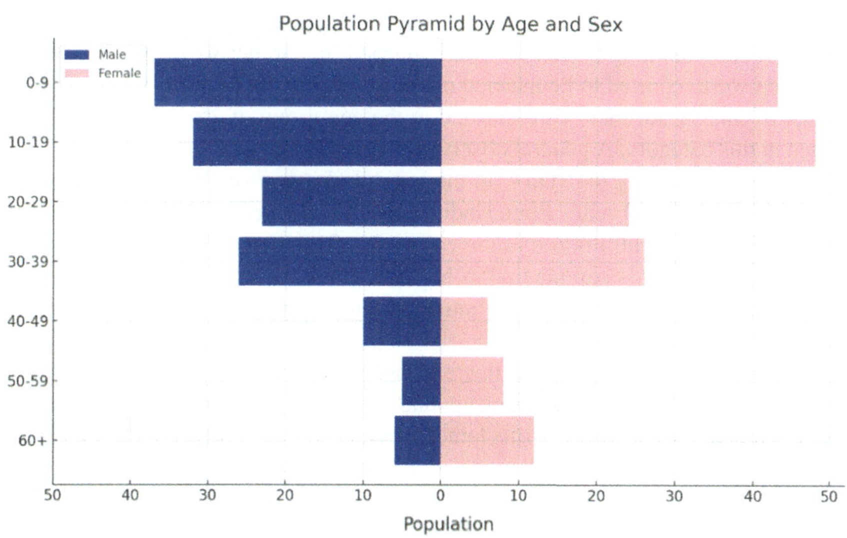

The data from the table could be understood and analysed from the above pyramid for a more easily understood purpose. One notices the minimised number of men is an outcome of occupational tragedy, caste-based labour, and how living in a menial job demands losing your senses and capacity to carry the filthy waste. So, alcohol is a temporary remedy to the workers, but ultimately it turns out to be a means to early death or increased morbidity.

2.10 Marriage and Kinship Structures

Hadi Cooli has an average of six people per household, and one can notice that several houses are built close to each other; their walls are shared, but most houses are small in size and structure. Two brothers, Krishna and Binod Hadi, have a typical house, but they cook and live as two separate households after both brothers married. Similarly, Ramesh's son lives in one house, while another resident, Lila Devi, and her husband have a separate house since she is married to another person. Lila's father gifted a house to his daughter and son-in-law, Ramesh Hadi, to establish their household within Hadi Cooli, so they do possess a house, but it is small. One can find a closely knit community of sentiments and emotions among them. They share a close familial interaction. As mentioned in the previous discussion, Dhulia resides with her youngest son, Jhomda Hadi, who works at a private office in Bokaro Township. Her other sons, Komda Hadi and Somda Hadi, live in a separate house with their wives and children, but near their compound walls. Pintu and Goli's houses are located next to each other and share a common wall.

In a significant observation of the data, there are 160 married individuals, of whom 19 are widows and one woman is a divorcee. In the age group of 10–17 years, nine males and the remaining 57 females were married. In another segment, 48 men and 23 women were reported to be married in the age bracket of 18 to 24 years. A small number of three Hadi men were married in the age group of 25–31 years, with no women in this category. A general observation about age at marriage indicates that it tends to be early among the Hadis. In the last decade, the average marriage age has increased from 15 to 20 years. Pintu Hadi recalls that his marriage was not his choice, as his parents forced him into it at the early age of 12. It was when "I was playing with peers, and then his parents told us how pretty the bride was. Where would you get such a beautiful wife in Hadis?" Pintu mentioned that his entire family, including his father, liked the bride.

Marrying at an early age invites the responsibility of running a household in Hadi Cooli. Fulfilling the wife's basic needs and managing an independent household increased his tensions. However, his father provided support and promised that Pintu would not face unforeseen challenges after marriage, and it turned out exactly as promised. Today, he lives as a happy head of the household and recalls his past as a mere coincidence. Pintu married his eldest son at 24 and had to arrange marriages for his other sons and daughter. During field days, his son Vishnu was 19 years old, while his daughter was just 1 year younger, at 18, and performed household

Table 2.3 Age at marriage of ever-married Hadis

Age Sex	10–17	18–24	25–31	Total
Male	9	48	3	60
Female	57	23	–	80
Grand total	66	71	3	140

chores. Vishnu worked as a sweeper and earned a monthly income, contributing to a seemingly happy family with three earners. Pintu and his wife take a deep breath and feel satisfied, stating that he did not make his children marry early, against their wishes. It is unwise to arrange a marriage before age 20, and he believes that even children could find a partner. They would be fine in such situations and help by following all rituals and ceremonies for the children's happiness. It is also a social fact that most boys in Hadi Cooli do not express a desire to marry at a young age or get engaged to settle down as heads of households. Delayed marriage minimises the burden on the youth and extends their enjoyment of youth, which most Hadis cherish after work each day. They spend leisure hours sitting with friends in nearby tea stalls as a routine. Table 2.3 presents significant sociological data.

Many women and a few men were married before age 18. Fifty-seven women got married between the ages of ten and seventeen, which indicates a higher number of women are pushed into marital life, unlike men, of whom only nine are in the age group of 10–17 years. Notably, women at Hadi Cooli are fewer in the paid workforce than men, which offers a unique insight into any sweeping community being studied nationwide. As shown in other studies, many women work as menial and sanitation workers in urban municipal corporations, unlike the sanitation workers in Chas town, where men primarily dominate.

2.11 Household Compositions among Hadis

Hadi Cooli has a mixed household composition. Hadis' household configuration indicates that they possess both complex and straightforward structures. Of the 51 households, 23 are complex, and 28 are simple. Jhomda, his wife, and their child live with his mother, while two of his brothers reside separately with their wives and children. Jhomda said, "I wanted to live with my mother, whereas my elders did not prefer that. They wanted to live separately so their households could have better lives, good food, and minimal daily complaints." Previous sociological literature shows that low-caste families tend to prefer and practice nuclear households for various socio-cultural reasons. Major studies on Indian households indicate that people from lower castes within the Hindu caste structure prefer living in simple households after marriage (Shah 1996), and this is a common social process. It is typical for married sons to move into separate households. The number of joint and simple households is almost equal. The younger generations express a desire to live separately as each contributes to the household through their earnings. It is also true that lower castes often prefer moving into separate households after marriage, a widespread practice. An eminent sociologist, A. M. Shah, who wrote extensively on the nature and concerns of family and households in India, mentions, "The emphasis on joint household was greater among higher castes and classes, who formed a small section of society, than among lower castes and classes, who constituted the vast majority of the population" (Shah 1998, p. 3). Thus, it becomes evident that Hadis' preference for simple households is not uncommon but aligns with the experiences of many other lower castes as described in A. M. Shah's literature. Individual earnings, the choice to spend on their wives and children, food preferences, and types of recipes become significant for Hadis to the extent that they transition into simple households after marriage. However, Hadi Cooli's limited land and a lack of savings or funds restrict them from moving immediately into new houses. Some received support from their in-law families to construct a new house post-marriage, while parents and other relatives also contributed.

However, Jhomda feels differently. He views a joint household as a loving family supporting children and the elderly. Similarly, elders like Ramesh and Somesh believe that the joint household offers a better environment for helping one another and passing down traditional and family values to newer generations. Nonetheless, the shift towards simpler households has begun and is gaining acceptance among the Hadi youth. There is an increasing number of simple households, around 27, which surpass joint households, and some individuals are on the verge of transitioning to simple households. This change is also being viewed in light of the existing occupational structure within Hadis. Their earnings are limited, and a considerable number of women serve as homemakers compared to 30 women working outside, either in homes or private offices. Limited income makes managing food, health, and other expenses challenging. Therefore, once married, Hadi individuals believe settling into simple households is better than living with others.

The most important reference could be found in a volume titled The Family in India, wherein A M Shah wrote, "Urbanisation does not lead to disintegration but only to the transformation of the joint family. Social Scientists do not, however, question the belief that the joint family was always a characteristic of village India (Shah 1998, p. 52)." Hadis live in an urban area and have joint and straightforward households. There are also households with an elder widow sister who lives with her brother, his wife, and unmarried children; a widow living with her married daughter, her son-in-law, and unmarried grandchildren. The nature of the household in Hadi Cooli is more than joint and straightforward. It was also found that Hadis had fewer choices to leave Hadi Cooli due to better chances of employability in Chas. Chas Municipal Corporation provides regular jobs, though the wages are low. Even such sweeping jobs are unavailable in other small towns in Bokaro and Dhanbad.

The Gair Majoorwa Zameen has nearly been reduced and is occupied by dominant families and construction builders. Although little is known about the area, such land is primarily sold to migrants at a higher price. Even land on both sides of Joria (now a Nala/ drain) has been encroached upon. Hadi Cooli has become very congested over the past 15 years. Consequently, new households have to reside in a small room. Migrating out is feared to jeopardise their work. Travelling from the outskirts of the town is not easy due to high commuting costs. Moving to another town would mean Hadis must live far away from their kin. In this context, Hadis continue to live and manage their lives within Hadi Cooli, even though the number of households in the area has increased over the last 15 years. A household census study in 2001 recorded 38 households. Currently, there are 51 households in Hadi Cooli. The composition of households has changed among the Hadis mainly due to marriage, death, and partition, which has likely contributed to the increase in the number of households in Hadi Cooli.

The residential settlement of this community is geographically quite complex since the space is very congested and cramped. The doors are small in height, not higher than 5 feet, and shorter than the height of most of the household's adult members. Electricity and water are not available to all. A few families have made proper electricity connections, but most households connect to the primary electric wire and use unauthorised electricity without paying any charges. Except for a few houses, most are kutcha and semi-pucca houses. Kutcha houses are thatched with soil, mud, and bamboo sticks; the flooring is soft mud. The semi-pucca houses have cemented walls and thatched roofs. These houses also use Khapra (a kind of sun-baked, earthen, rounded piece, but not bricks) on the roof. The sizes of rooms are 10/10 feet or smaller. One room with a veranda is a typical house structure, while only a few have two rooms with a Veranda. The veranda is a half-covered area without doors, and it is used for cooking purposes. After cooking and eating, elders and children rest and sleep in this area. Generally, old people and children do not need privacy in the house. This type of house requires annual repairs after the monsoon because the rains damage the roofs of the Khapra (earthen tiles). Many households have to spare for such expenses; hence, monsoon is not a grace of God; instead, it is a curse in their life.

Hadis's residential surroundings are very congested. The housing pattern lacks proper ventilation, preventing fresh air from circulating in most houses in Hadi Cooli. The absence of proper ventilation also increases the risk of illness. Hadis lived in small rooms where all members had to share the space, which felt claustrophobic. Except for three households (discussed in this chapter), all others lack access to basic sanitation, such as a toilet. People are not particularly concerned about the health problems arising within the community due to this lack of basic sanitation. Cultural resources in the form of information and awareness might pose health threats, whether because of congested housing, open defecation, filthy residential surroundings, etc., or by increasing the vulnerability of Hadis. The resources available and accessible to community members for dealing with such illnesses also influence toilet usage practices.

During summer, they can sleep without fearing weather disturbances, but not in winter or monsoon. It was found that the doors of a few houses were very weak. When I enquired about the same, they said that they do not feel troubled and do not have a fear of theft. Lakhicharan, 60 years old, Hadi works as a sweeper in ICICI Bank, located on the Bye Pass Road of Chas, says, "*Hamni ghar mein kutch rakhbe tabai na dar lagtai, Kahee khateer kutchu rahbe na kare khae khatir, to kahil dimaag kharab kare ke. Kee chori bhatae, kutchu hayye nakhai.* (We do not have anything at home; at times, there is no foodgrain for more than a day, why bother and fear theft.)" Nevertheless, theft is not uncommon. The financial crunch encourages the unemployed youth to steal from vulnerable houses. During the summer, theft is reported in several other Mohalla houses, but not in Hadi Cooli. Hadis feel that a few other caste groups from Bauri and Shakaro do it. Hadi never attempted to steal. The other castes, such as Shakaro and Muslims, also believe it is mainly from Bauri and other castes, but not from Hadis.

Within Ghoda Gotra as practised among Hadis of Hadi Cooli, their own kinship network is structured, and they rely on blood ties, marriage, or adoption, and it is outside the Cooli. In common sense, marriages are not arranged within Ghoda Gotra, as the choice of bride/groom is practised with matches outside the Hadi Cooli that follows Gotra. Such practices are called gotra exogamy, a social practice of not marrying within the same gotra. The rule of avoidance for marriages is limited to the paternal lineage, where they avoid forming marriage alliances in a family where there is a known lineage. To clarify, the groom from Ghoda Gotra cannot marry a woman from the same gotra if, in any distant traces and knowledge, she is an offspring of a male related to another male line within the traceable depth of Ghoda Gotra. In the vicinity of the town of Chas and the entire district, families do marry with another family but with a known gotra. Ramesh Hadi was also married in the Hadis, but from another lineage. Hadis also do not marry the progeny of their father's siblings. The practice of avoidance of the above nature is followed and checked until traceable generations by Hadis are found in the male line. A complex kinship network shows that Hadis' marriage alliances are arranged within Maghaya Hadi, among all the four sub-castes mentioned in previous sections of this chapter.

Married women reside in various places, small towns, and localities in the earlier mentioned districts of the Jharkhand state. Dhanbad has the highest number of married women's natal localities among the ever-married women, Hadi Cooli. The localities from where these women came to Hadi Cooli after their wedding are as follows: Baghmara, Bhaga, Bank More, Chirkunda, Steel Gate, Katrasgarh, Kola Kusma, Tetul Mari, Kenduadih, Govindpur, Mugma, Hill Colony, Jharia, Digwadih, Mahuda, Bhooli, Nonudih, Gomoh, Topchachi, Lodna, Kesar Gaddha-Dughda, Sijhua, and Maheshpur-Katras. All these areas fall in the Dhanbad district. Three women are from the Ranchi district. Those three localities are Doranda, Dom toli, and Ranchi town. There are only two women's natal homes in the Giridih district, Dumri and Girjhingri. The districts of Hazaribagh and Ramgarh have each sent one woman to Hadi Cooli upon marriage. The places are Harinagar and Ramgarh Cantonment, respectively. Purulia district also shows only one such natal locality from which a married Hadi woman came upon marriage with Raju Hadi, Adra. Raju Hadi has a house and is settled and working in Adra. Out of nine natal localities from where the women have been married to men in Hadi Cooli, 45 have come from various localities of Dhanbad, and the same is mentioned above. Six of the 45 weddings occurred in Chas, three at three locations, namely Tupkadih, Sector 4 in Bokaro Steel City, and

Bermo, and others at Hazaribagh, Giridih, and Ramgarh in Jharkhand. The wedding also took place in Adra in the Purulia district.

As mentioned above, marriages are preferred within the sub-caste of Magahya Hadi. Ramesh Hadi says, "We will not get bride or groom outside the Maghaya Hadis." However, marriage alliances occur within the Hadis living in the above-mentioned districts. However, inter-caste marriages are not permitted, even between different Scheduled Castes. The elders recorded that the alliances are arranged with the help of a kin member or a caste member. Once the information about a bride or groom is attained, the household elders and at least one elder from Hadi Cooli initiate the proposal. *Buddha-Purna* (the senior-most elders) has an important say in bringing or giving away a bride. The younger family members respect the suggestions and decisions of the *Buddha-Purna.*

A few parents, like Goli Hadi and Pintu, added, "We also ask our children if they have any choices unknown to parents." However, when asked whether they give similar choices to their daughters, both Pintu and Goli did not have a clear explanation. Goli opines that daughters should also be given a choice while getting married. My daughter should be able to discuss the alliances in the household, but this is not agreed upon among most Hadis. Marriage is considered a family affair, and an individual's choice is only a polite formality. Elders in the community help inquire about the prospective alliances and the necessary details about the family with whom the alliance is contemplated. For instance, Haribabu has three sons and one daughter. All the sons married in different families outside Hadi Cooli. However, they lived in their paternal residence, and the married daughters resided either in the houses given to them by their father or constructed their own houses in the Hadi Cooli area itself. The father gave the land. However, these daughters are married and settled after marriage in a new household within Hadi Cooli; the rest live with their husbands at their respective paternal houses.

Many Hadis have re-constructed their old houses with financial help and support through the Chas Block Development Office (BDO). Following the broader guidelines of the National Housing Policy and Habitat Policy, 1988, there is a possibility of accruing benefits that Hadis can explore. It is a programme of the government of India for the poor population, especially the houseless and the inadequately housed, improving the settlements' environment through providing sanitation, drinking water, etc., while giving preference to SCs and other weaker social groups. The housing condition tells the conditions of a cooli of Hadis. There is a decreasing space concerning the population of Hadis. Three toilets were constructed in Hadi Cooli under the government-funded schemes for poor households. These toilets are mostly unused. There is no water facility in them, and most women do not prefer to use them. They also prefer going out in the open for defecation. Most of the Hadis have not been able to approach the BDO office to access Indira Awas Yojana (IAY) benefits.

2.12 Transforming Traditions in an Industrialising Landscape

Since the 1970s, migrants have had opportunities to earn and live with decent earnings compared to Hadis. Thousands of people migrated to Bokaro from different districts of Bihar, namely Samastipur, Darbhanga, Ara, Chapra, Siwan, Sasaram, Muzaffarpur, and the district of Balia from Uttar Pradesh in the 1980s. This migration from other states resulted in the expansion of the urban population of Chas and its outskirts.

The newly formed Jharkhand state did not prosper despite its rich natural and human resources. Although the development and growth of the Bokaro Steel City have helped in the economic development of undivided Bihar, presently in Jharkhand, the gains have hardly reached the Hadis. Bokaro is well-developed, and it can be gleaned from the data that out of a total of 280,000 residential quarters constructed in the entire country at various sites for the workers in the public sector projects, no less than 60,000 are located in this small region alone (Census of India 2001). These houses are built on the steel plant premises and allotted to the employees in the plant. However, the benefits have accrued to the dominant sections, not the Hadis or other dalits.

The number of migrants settled in Bokaro district, over time, assumed political power in the erstwhile Dhanbad district in the 1970s after the nationalisation of coal mines and the establishment of BSL. However, those employed in BSL have ideally little stake in the political power in the Bokaro district. Sudhir Hadi, a resident of Mahuar Basti, elaborated, "The Majority of the employees are migrants from other districts and states. However, the dominance in the steel plant continues in the hands of those select migrants who are fewer in number but most influential. The present Legislative Assembly (MLA) member from Bokaro, Samresh Singh, has been elected in the State Assembly elections. Due to his political clout in the state Assembly, he runs a few Bokaro gymnasiums, clubs, and a hotel. Most importantly, all the coal, BSL, and mafia contractors sit in his office. They have close informal ties and help each other whenever in need. Employees of the steel plant are less concerned, but they are made to support the political and powerful clout of Samresh Singh in the township area. However, the natives like Hadi, Bauri, Shakaro, etc. have less say in demanding their rightful benefits," in an interview in January 2011. Sudhir is an employee of Bokaro Steel Plant and an active member working to uplift Hadis in the state through his NGO.

The emergence of the erstwhile district, Dhanbad, as the coal capital of India, including several other prominent industries in the same period, attracted many migrants. The dominance of migrants is reflected in the places named after the districts from which they migrated. Most have come from Bihar and Uttar Pradesh, and this is reflected in the turns and roundabouts in the districts being named Arah More, Chhapra More, Balia More, and Bhojpur More, etc. ('*More*' refers to the road junctions or roundabouts on the road). Several leaders from these districts were strong regional leaders, namely, Suraj Dev Singh in Dhanbad and, at present, Samresh Singh in Bokaro. In residential colonies with a sizeable particular population, the areas around them are named after the migrants' language and region. For instance; people from Bhojpur region of Bihar have named an area as Bhojpur Colony, Gujaraits have named one locality as Gujarat Colony, Chapra is a district of Bihar and people in one area called it Chapra Colony, and so a few other residential areas are being named by the dominant population from one region, Rampur Colony, Marwari Colony. Further, the above-mentioned colonies also show the pluralist nature of the population settled in the town.

Chas, once a green and agricultural land, presently has no greenery, is filled with small-scale industries, and has recently experienced an enormous housing expansion. This reminds one of the meanings of the word 'Chas', which is derived from the Bengali dialect, meaning 'agriculture'. Most of the land in Chas was agricultural until the Bokaro Steel Plant came into existence. In the year 1972, Chas was converted into a Scheduled (Adhisucheet) Area to protect the life and traditional livelihood of Dalits (SCS) and other backwards populations, including a few tribes. In this backdrop, the steel plant was shifted from Chas town to the present location of the Bokaro area. In 1984, following the recommendation of the Tadarth Committee, Chas Municipality was formed as an independent civic body, and Chas became a town in 1984, parting away from MADA.

During this period, the establishment of Bokaro Steel Plant Limited (BSL) accelerated the population increase, and consequently, the need for drinking water, sanitation, hygiene, electricity, and street roads increased. Chas witnessed a rise in population from 5000 in 1951 to 393,805 in 2001, thus exerting immense pressure on the basic services and amenities. Lastly, Chas town was again upgraded into a Municipal Council on June 26, 2010, followed by urban ward elections. These elections were held almost 30 years after the last election. Elections had not taken place under the 74th Amendment Act of 1992. The absence of local wards affected the town's developmental activities (Chas Master Plan 2027).

The sanitation and sewerage conditions have not improved compared to other areas in town, and the availability of drinking water has worsened. The Hadis live a segregated social life, even though Hadi Cooli is in the middle of the town and market area. More often than not, they have a community life and continue carrying out their traditional occupation. The region's population has increased more than fourfold during the past 50 years, primarily due to immigration. Sengupta's (1980) paper gives the following data on migrant workers who pushed the native labourers to the margins in the 1950s before the Bokaro Steel Plant came up. The linguistic range of the migrants listed in Table 2.4 reveals this.

Sengupta (1980) further reports that Bauris have historically been labourers in agricultural farms and later at various mines and small-scale industries, so that they would comprise the largest labouring community in the state. A miserable lifeworld began for Hadis in the 70s and 80s due to the nationalisation of entire coal mines by the government of India. Within a week of the nationalisation of all coal mines in the year 1971, an estimate shows 50,000 miners lost jobs and they replaced by migrant populations largely arriving from certain pockets of Bihar and Uttar Pradesh (Bhojpuri-speaking region, a colloquial dialect in Bihar and eastern

Table 2.4 Linguistic decadal growth in the erstwhile district dhanbad

Languages	1951	1961	Linguistic decadal growth		
			Linguistic region	1951	1961
Punjabi	30,000	57,000	Bengalis	7.2	9.7
Marwari	8000	22,000	Oriya	3.1	3.0
Gujarati	8000	36,000	Bhojpuri		113,000
Tamil	6000	15,000	Maithili		16,000
			Chattisgarhi		9000
			Bilaspuri		2000

Source: Sengupta (1980)

UP). Sengupta writes that thousands of postal telegrams were being dispatched from Dhanbad district by migrant workers to the people in their homeland to Arrah-Balia-Chhapra districts, and informed that ample jobs were available. During the Emergency (1975–1977), thousands of workers were retrenched, and they were all from Adivasis or Harijans' communities (ibid, 1980) due to the political pressure of upper-caste groups hailing from the Balia district. They were removed due to the growing political influence of migrants in the coal region.

In contrast, many Hadis have not migrated elsewhere and have been further confined to their small localities. Unlike others, they have not entered any new occupation in repairing, driving, etc. Most were not aware and fully occupied in the traditional work they were already into. The other reason that a few Hadi got a job in BSL was those who were aware of the upcoming opportunities: Somesh, Pintu, Shailender, Jagdish, and Ramesh in the first generation, who got it. Some others were already working in MADA and later shifted to the Chas Municipal Corporation.

2.13 Civic Entitlements and Deprivation

The lives of Hadis are not restrained by economic constraints only, but the absence of basic amenities has also made their lives full of difficulties. Everyday struggles are high because small ponds (usually this region had several ponds in every few kilometres) located near Hadi Cooli were just three, and these get dried up in the peak of summer months. It is worth noting that families depend on such ponds for daily bathing, cleaning clothes, and sometimes washing utensils. In 2009, the government installed a hand pump (deep boring) near Chas bus stand. This is a significant drinking water source for most of the households of Hadis living near the Chas bus stand and market. Another hand pump is near *Sitala Mata* temple, which helps some families of Bauri Cooli and Shakara Cooli. None of the Hadi households has access to a personal borehole pump.

A few Bauris have been able to get a personal handpump. "It is very costly to establish your hand pump," said Ratu Bauri, a resident of Bauri Cooli. Boring would cost at least Rs. 30,000 rupees if groundwater is found. After collecting water, women have to prepare food for their families. One young girl at the water supply point said she must fetch water early in the morning. The water supply is available for less than an hour, and sometimes it stops before everyone gets the water in their pots and buckets. The crisis of water also affects the Hadis' daughters. A Hadi girl said she wanted to join the school in her locality. However, she regrets because, "*Hamra timen milo hai bihan ke school jael kai, Jhaadoo pocha, Paniya bharail wala aisan kaam ho kee chodain pariye hum.* (I do not get time in the morning, and I have to fetch water, which leaves little time to reach school in the morning.)" The girls are kept busy helping with domestic chores at home. A few women opined that education would not be important to their daughters' lives. She must take care of her husband and household activities whenever she marries. Some women want to educate their children.

Water pipelines are only available in some Mohallas. A few taps are installed at familiar places, but most houses of other Mohallas have borewells to fulfil basic needs. Hadis manage their water needs at the Mahabir Chowk area provided by the corporation, which is meant for all. The supply to a location closer to Hadi Cooli could make their lives easier. A few of them say that soon, there is a proposal to have a borewell to help all Hadis.

Doctor Vimal Singh said, "I must begin the initial treatment. Our understanding is largely based on the way patients explain the illness. The drinking water in the Chas area is hard due to the high mineral content in the ground. In some areas of Chas, underground water is pure and tastes good. However, the unhygienic drinking water causes frequent stomach infections among Hadis and Bauris. Both the caste groups are largely dependent on the public taps and borewells. The public tap located near the bus stand is in the risky zone. It is in a crowded market area, and the road has the heaviest traffic. Present-day sewerage is a natural stream, Joria, which gradually becomes a drain. The sewerage passes from the bus stand at one point, making it riskier. Hence, the chances of water mixing with the drain or sewage are high."

The water tap was probably above the drainage, but gradually touched the drain side. The filth and unhygienic location may affect the potable water. Women keep their pots and buckets and stand in queues for hours. Vimal Singh describes water fetching and the location of the water tap as unhygienic, which is an additional disadvantage for Hadis. The deficiency in providing water to Hadis is severe compared to other SCs. For instance, Bauris have hand pumps, one well, and a public tap near the Sitala Mata Mandir and Hanuman Mandir at Bauri Cooli. It is known that there are a couple of houses with private hand pumps. Most Bauris are better off than Hadis. No comparison can be drawn between the two castes, although both belong to SCs. Nevertheless, the drainage is similar in all the localities of the neighbourhood, as they are neither maintained nor cleaned. Most drains are open, and that invites waste to be dumped in them. In this context, Hadi Cooli is slightly better as the drain passing through it is cleaned.

Electricity supply is not regular for the Hadis, giving rise to illegal connections. With the help of an electrician in the locality, a few households connect electric wires to electric poles. No metre is installed, and when a government official arrives for a visit, these illegal connections are removed. After the Assembly and Ward elections in 2010, the streets improved. Hadis felt happy to see the lamp post installed at the entry point to Hadi Cooli in 2010. The lack of toilets, a hygienic environment, and a garbage disposal system, however, do not make it to the Hadis complaint list. Their priority is to find better working conditions and permanent employment. The lack of basic amenities is reflected in the deficiency of social interaction between castes. The notion of filth and pollution is attributed to Hadi Cooli, which emerges as a barrier between the Hadis and non-Hadis. Primarily, social interaction is limited to interactions within the caste groups, and inter-caste relations are less observed. This attribute is learnt during early socialisation by children. The youth establish social interaction when they grow up, but it is at its lowest during childhood. The school is located outside Hadi Cooli and it caters to Bauri and Muslim *Mohalla*s.

2.14 Sanitation and Sewage Conditions

Though Chas was recently recognised as one of the cleanest municipalities in eastern India, looking at the sanitation and sewage system, apathy and miserable living conditions could be seen and experienced. On top of this, Hadi Cooli is the worst affected in these indicators. Late in August 2011, Ramesh Hadi said, "How can there be such an insensitive government even though the head of the state is represented by a native tribal, Shibu Soren, then Chief Minister of the State, one of the most neglected social groups of Indian society?" A daily newspaper, long ago a popular English daily, the Times of India, had reported that there was no single sewage treatment plant (STP) in the city and probably in the entire state (TOI, 11 May 2011). Though in another report of the Jharkhand Municipal Act (2011), three STPS were run by the state administrative units of the concerned districts. It is an essential mission to achieve total sanitation and a hygienic environment under the MDGS globally, and currently, the SDGS emphasise this effort. As per the discussion in the field in the year 2011 with the district officials of various departments, it was found that the untreated water goes directly into the river flowing through the state, and most of the water bodies are also getting contaminated, resulting in serious health hazards. During summer, the river Garga would almost dry up, and the drainage water would contaminate the remaining water. Even underground water is not good, and the water table is depleting yearly. His lifeworld continues to be in destitute and persistent poverty that does not allow it to break its cycle.

In the areas of town, an alarming increase in groundwater contamination has affected the primary drinking water sources through borewells, wells, manual hand pumps, and tube wells available in the vicinity of the settlement, *and* Hadis use these. One report related to the total sanitation campaign, Jharkhand state lagged at the bottommost states against the national average of sixty-three (63) per cent; further, only 41% of households in the state have toilets at their houses. The absence of sanitation and toilets adversely affects everyone, especially women. One of the documents of the World Health Organization (2009) cited from Rebecca (2008,

p. 172), "Sanitation is a cornerstone of public health," pointed out. The lack of toilets not only challenges the dignity of Hadi women but also exposes them to bacterial infections. A local doctor recalls recurring diarrheal infections among the children and women in Hadi Cooli. The greater concerns are the growing contamination of soil and water due to open defecation on the Joria side, which was earlier a water stream and is now a Nala. Hadis use *Joria* water for several domestic purposes. In Hadi Coooli, five households have private toilets at their houses. Ramesh Hadi, Goli Hadi, Krishna Hadi, Binod Hadi, and Pintu Hadi have constructed private toilets in their houses. The government schemes supported three. But these are used only by females of the house, and the males still practice open defecation. At times, females also go out for defecation. The data related to Open Defecation (ODF) has a complex nature of concerns and demands more socio-cultural and political wills required; urban spaces must be ODF-free at any cost, so latrine facilities at every household across caste and class have to be made available.

The interviews with several Hadis in the field revealed that water is filthy and smells during summer. Nevertheless, Hadis have no alternative and wash their bedsheets and *gendra* (home-made bedding) in the *Joria*; this increases the vulnerability to infections. However, the data does not prove the causal correlation of disease in Hadis with the utilisation of Joria water, and a different specific study is needed to examine the same. Certainly, recurring infectious diseases, namely stomach disorders/gastroenteritis and related illnesses, bronchitis incidence among children, and regular reporting of fever and cold, give an idea of infection due to such unhygienic surroundings.

During monsoon, while I was in the field, most doctors in the neighbourhood reported that the season was perfect. A good season for the doctor is when there are high incidences of illness and, in turn, many patients report to the health care providers at clinics. This year, the market is better than in the past few years. It was the same year when an unknown fever similar to chikungunya was reported in 2009 and 2010. Hence, the turnout of patients was very high in most clinics. Dr. Jamal, who runs an honest practice in another neighbourhood locality, shared with the researcher about his medical practice. He reported, "I have seen several endemic seasons when many patients fall ill with fever, skin irritation, and diarrheal diseases. I would not even get four to six hours of regular sleep, but you ask my patients, I never overcharged. Precisely for these ethical practices, I have the highest number of patients across the *Mohalla* and even a few patients visit from Bauri Cooli, but in Hadi Cooli" (Photograph 2.1).

Photograph 2.2 shows the entrance of Krishna Hari's house. They belong to the lineage of the first settlers in Hadi Cooli. Hence, they could have managed a better house, but the rooms were tiny and congested. However, there is one corner in the courtyard allocated for prayers. There is a large number of households that have some familial association with the first settlers. It is a well-known fact in the entire region that Bokaro was the most industrialised urban centre in the 1970s, but gradually, the city lost its growth, unlike Jamshedpur. In an interview in the field, Sunil Kumar, Bokaro's Deputy Commissioner (DC), especially reiterated that Bokaro stands out for its greenery in the entire state. However, the adjoining town of Chas is in contrast

Photograph 2.1 Krishna Hari's two private flush toilet at hadi cooli, Chas. Source: Ziyauddin (2022)

to it. Before its separation from Bihar, Bokaro was one of the well-planned and developed towns. However, the development is confined to the steel plant's industrial and civic areas, leaving Chas out. In its May 22, 2011, (Hindi edition), Hindustan Daily reports that Chas has transformed from a scheduled area into a municipality and is presently a Municipal Corporation. After four decades of establishing the Bokaro Steel Plant, nothing has changed except the growing uneven population in Chas town. The adverse effect of increasing population is seen in Chas town and not in the Bokaro Steel Plant residential area. The town never had proper planning for its future growth, worsening the sanitation situation for the people, especially in poor communities. The sanitation workers' lives looked like a story of persistent denial of dignity and respect in their jobs for decades until 2023, when another round of field visits was made.

2.15 Life in Hadi Cooli

Life in Hadi Cooli is a hustle and bustle early in the morning. Morning is when all the residents wake up and start getting ready to go to work while it is still dark. Women approach Joria stream and the pond near Chas main road to ease them. The 30 working women have to take extra pains in the morning hours before they go to their jobs. A delay would make it difficult for them to start their everyday work. Men

Photograph 2.2 House of krishna hadi. Source: Ziyauddin (2022)

can go for defecation even a little later. However, women cannot delay this as after this, they have to fetch water from a tap installed by the Municipal Corporation near Mahabir Chowk off Hadi Cooli. Most men, too, would try to attend to their work as early as possible.

Men rush to work either after having breakfast or, at times, without breakfast if food is not ready early in the morning. For instance, Roshan Devi's husband and sons walk for 15 to 20 minutes every morning to reach their workplaces, municipalities, banks, and other offices before eight or nine o'clock. Her husband, Somesh Hadi, goes last in the morning because his office opens late, i.e. by 8:30 A.M. Her elder son, Mangru Hadi, aged 33, works at Chas Nagarpalika. He is the first in the house to begin work every morning. This household is a joint one, which makes it easier to finish the household chores collectively, adds Lakhi Devi, Roshan Devi's daughter-in-law. She says that Lakhi Devi and her mother-in-law try to prepare breakfast in the morning, but it is impossible every day. Our men go without food in winter, but the food is ready when they return after work by 1 PM or later. For this reason, sometimes Roshan Devi cooks a little extra food for dinner, and the leftovers are kept for breakfast.

Raman Hadi, the youngest son, does not come home for lunch but generally eats a heavy breakfast like rice and daal, sometimes with *tiyan sabzi* (vegetable). She sweeps rooms and the front of the house before joining an army of womenfolk to fetch water. Finding a healthy adult male member in the house is difficult in the early morning hours. Everyone in the household goes out to work for their livelihood, confined to sweeping, cleaning latrines and drainages, cleaning and collecting garbage from households in different localities, offices, banks, hotels, and hospitals, among others. Garbage rickshaws in the streets and roads during the first half of the day in most of the localities of Chas town are very noticeable. The significant respite for Hadis is that the filth and dirt are not carried as headload. The work in Municipal corporation localities ends by noon or early afternoon. Hence, most Hadis go back home for lunch at 1 PM. If the work remains, they go back to attend to it. This happens mainly to those workers who clean drains and collect garbage from roads.

However, those not working with Nagarpalika (29) cannot be late. Forty men out of 75 (male) Hadi work in Nagarpalika. For instance, Jhomda Hadi goes late to the city centre market in Sector 4 of Bokaro Steel City, the main shopping centre. He lives with his wife and mother in Hadi Cooli. His day begins early. In recent years, both men and women have had to get up before sunrise to go for defecation. His office starts at 10 a.m. It is a private office called Tata Indicom retail shop where one gets mobile recharges, data cards, sales, and purchases of phones, etc. However, unlike others, he has to work till 5:30 pm. This difference does not make his life difficult. Instead, he is pretty happy after working for a few years.

All this shows Hadi's life pattern in Hadi Cooli. However, the day's life as a community in Hadi Cooli does not end after work is over; instead, it begins after work. The freedom and relief after the day's cleaning work are observed in the afternoon. A few elders say this has changed over the past 20 years. Hadis would indulge in alcohol after finishing their work, and some people sometimes drink even before they go to their jobs. Haribabu compares his present with his past life. Several of his community members were alcoholics. Haribabu is 62. He is still active and says, "Sometimes, I even drink with friends in the neighbourhood. It is easier to find liquor in our neighbourhood than anything else. But I never got addicted to alcohol. The credit also goes to my wife, Fudni Devi. She was a kind and mature person who explained household problems and our kids' future to me. I find a difference between me and my peers in Hadis. Alcohol has affected the health of most of my friends and taken a great toll. Even today, I work in two shops near the main market. I earn 4000 a month by sweeping and cleaning the offices and toilets. I cannot do heavy work due to my age. However, my wife and I manage our everyday needs on this amount." Alcohol has affected the lives of most Hadis who drink regularly. A few could avoid alcohol dependency and feel happy about it.

2.15.1 Financial Conditions of Hadis

The financial condition of Hadis has improved compared to the past due to the town's expansion. However, they continue with the menial occupation due to the apathy of the state machinery in the town. People along the main road, Bye Pass Road, and several other localities have opened shops and rented their front rooms to others. This advantage is not for Hadis. Ramesh narrates, "We are confined to the insides and not the main road" except his son, Kaali Hadi.,[1] who has a small grocery shop at one end of Hadi Cooli on the entrance side of the Cooli. Several Hadis reported that they hoped they would get permanent jobs after the state's separation from the erstwhile state of Bihar in 2000 AD. Nevertheless, this did not happen even when the new state was created to uplift its citizens.

In addition, there is an ingrained fear among Hadis of availing benefits from various government-sponsored schemes available to them. I observed that Hadis feared that the local officers in the block might scold them and frame them in some fraud cases if they (Hadis) tried to benefit from these schemes. Lack of awareness about the available schemes among the poor further increases their deprivation. In some cases, it can also be seen that Hadis face much discrimination at various levels due to their occupation, which is why instead of availing the benefits from the schemes, Hadis find it easier to approach local money lenders and thus fall into their trap. A vicious cycle of less money leads to debt, and further inability to pay it back continues. Ranjeet Hadi narrates, "Due to an inability to save money at home, I borrowed money in the Sahu market from a moneylender. I had to buy gifts and clothes for one of my relatives' weddings. It was essential to attend and show my presence with gifts, etc. This compelled me to borrow 10,000, and it took me six months to return this amount to Sahuji."

Further, due to the existing social distance, it is rare to find government officials visiting the Hadi Cooli, "*Kahiil aithun ho, hamni ke Cooli choot lagain wokhni khatir. Yakon sarkari admi poochbai na karo hai, musalman Mohalla mein chail jeethu aar sob okhni se pooch le hathi, bujhain paariyo ho ke hamni ke kahi ke dand miloa ho" says Bajaj Hadi. "People in Chas do not enter Hadi Cooli due to Hadis's occupational background. People think that entering Hadi Cooli would pollute them. No government official cares to ask about us; they go to the Muslim Mohalla and get all our details, which are incorrect most of the time; we do not understand why we are being punished for so many years in society.*"

Most members reported apathy and a lack of concern from the officials towards their endless problems. Ashok narrates, "Nobody bothers to heed our request to provide a drinking water facility in the locality. Women and daughters spend hours fetching drinking water from Mahabir Chowk. The girls are the first people to be given this duty. They must stand for hours in the queue at the water tap to fill their pots. If there is a day when there is no water at home, men may go to other places located far away, i.e. *near Dharamshala More and on Chas Bye Pass Road or Chandra Talkies with plastic containers and fetch water. Men can use a cycle and collect water more quickly than women going that far and carrying it on their heads, as shown below in* Photograph 2.3.

In Photograph 2.3, a Hadi man, Lakhi Charan, collects drinking water in the afternoon near Chandra Talkies.

[1] Kaali Hadi is able to open a grocery shop in his own house. His house is located at the point where street starts inside Hadi Cooli and is slightly closer to main street towards Muslim *Mohalla*. They have two-storey houses. One portion is used to open the shop.

Photograph 2.3 Fetching water. Source: Ziyauddin (2022)

Women find it difficult to walk long distances, and one does not know exactly where water will be available. A few men use a cycle to go farther. In the afternoon, most Hadi men get free from work; hence, they find it easy to do this job. However, constant efforts and appeals are being made to the Municipal Corporation to provide a deep bore well in the Hadi Cooli locality. They say it is under consideration.

2.15.2 Food Habits

Most Hadis consume rice as their staple food. Rice can be served with curry, if available. Sometimes, some women cook extra rice for dinner that can be served for breakfast. This allows them to do other work early in the morning. However, several youths reported that they prefer to eat roti made of wheat flour with vegetables. Hadis visit the vegetable market near their residences in the main market area to search for cheaper vegetables. Elders reported that Madua was one of the staple foods until a few years ago, but it has gradually disappeared. Cheera Chas families used to cultivate Madua as their main crop. Currently, most of the land has been sold to housing construction companies, and there is no agricultural land available on the outskirts of Chas. Even villages have stopped using Madua; perhaps wheat has become a preferable option. Women are often asked to cook rotis during winter, which would be relished if they could afford chicken. A few Hadis mentioned that they cook liver as their non-vegetarian dish because chicken liver is sold at a lower price.

Most Hadis have an average income of less than 4500 rupees, except for those who work in steel plants and MADA. Expenditure on food is relatively high, followed by health and other expenses. Sometimes, men spend money on food and snacks outside, while women prefer to eat at home. Men enjoy eating with friends in the evening, drinking and snacking, and most like dining in hotels if they have some money. Women do not prefer to go to hotels for food, considering it unnecessary spending to eat outside.

The food habits of Hadis are primarily influenced by their ecological conditions, like any other community, where sufficient fish is available, and being non-vegetarian, they enjoy it. However, the drying ponds over the last 10 years have led to the disappearance of fish. Fishing was convenient in those ponds. During the monsoon, fish used to be available even in Joria, but now buying fish is costly. Fish sells for 120 rupees/kg, which is unaffordable for most Hadis. Machee (fish), rice, roti (wheat bread), and pulses form a significant part of Hadis' diet, though not regularly. Vegetables are consumed, and fruits are occasionally enjoyed. Alcoholic drinks, available in the market and the neighbouring locality, Bauri Cooli, are commonly consumed by them. Suresh Kumar Singh notes that Hadis preferred exceptional foods such as puri, lapsee (paste-like preparation), and pulao (biriyani of rice with fish or mutton) (Singh 1995, pp. 74–75). However, these dishes do not seem familiar in Hadi Cooli at present. With a few exceptions, Hadis of Chas do not fit Singh's description. Potatoes and some curry leaves are purchased at lower prices in the vegetable market. Most households cannot prepare roti, dal, and curry for a single meal, typically managing to cook just one curry along with rice or roti.

Generally, Maad-Bhaat (rice and soup) is the staple food among Hadis as it is cheaper and more economical. Ramesh Hadi shared in an interview that Madua (Raagi or finger millet) was popular among Jharkhand's rural and poor populations. It is cultivated during winter. He states, "Madua (Raagi or finger millet) is an extremely nutritious food. It serves as a staple food for the poor and many others. We did not own agricultural fields, but I saw Madua being cultivated. We used to receive Madua in exchange for some labour, a kind of barter arrangement. In our vicinity, Telgadia and Cheera Chas villagers cultivated Madua as their main crop annually. But today, you may not find it even in the Chas market. You can make roti from Madua flour or cook it in water, adding salt to taste. It was routine to eat Madua with onion and green chilli in the morning, sustaining almost the entire day. It is a heavy and nutritious food for the poor. A few Hadis recounted that our parents and elders served Madua to new mothers after childbirth, either with milk or salt and water. It is a very healthy diet for children and lactating mothers. You will soon see people returning to Madua, but they will not find anyone growing it anymore. Madua farming has significantly declined in our region." This change in the staple food has also impacted the health status of Hadis and women due to its nutritional value.

The tea stall at Bauri Cooli sells snacks. A spicy pakodi is preferred during evening alcohol consumption. Locally made alcoholic pouches are available in Bauri Cooli at varying prices, like rupees 10, 20, and 30. Occasionally, it is sold in reused plastic bottles, which also helps measure the quantity. Women find men's drinking habits burdensome; if drinking ceased, they could save some money. Men generally drink in the evening after returning from work.

2.15.3 Migration Hesitancy

Migration is rarely viewed as a positive idea by the elders of Hadi. Parents do not allow their children to migrate to other cities. There are specific reasons, such as who will care for the ageing parents back home if the new generations migrate. Similarly, how would the parents reach the migrated children in big cities in case of an accident or exigency? A few parents emphasised that they want to educate their kids in a good school, but it is difficult due to a lack of income. They could also send their children to other cities for better education and jobs. Goli is informed about the benefits of better education. He gives credit to Jhomda, who studied for a graduation and kept away from unclean jobs. The contradiction occurs due to various thoughts and perceptions among the Hadis.

Many of them find it easy to get a job as a sweeper. Sanjit Hadi said, "Even if we educate our children, what job will they get? There are many Bauris who have studied for a BA, but they are working as a labourer. We find our job and feed our dependents." This shows Hadis are somewhat overdependent on the available jobs in Nagarpalika and other avenues. Moreover, they do not want to migrate for socio-cultural reasons. Parents found that migration will make their children vulnerable, and so will the parents. Living together is most helpful in times of exigencies like health problems, attending to a relative in marriage, and death. The jobs in Chas have attracted several Hadis to settle down in Chas town from other places. But they have not moved out of Hadi Cooli in the last three decades. None of them wanted to migrate to other cities except for one or two members.

At present, none of the members of Hadi Cooli have migrated from their community to any of the country's big cities. Of course, there are immigrants in Hadi Cooli from the nearby towns and localities. Land records of Hadis show that some migration happened in the past, but parents do not want to send their children to faraway towns. Sanju Devi, wife of Suresh Hadi, hails from Dhanbad district. Her natal place is located at Hills Colony in Dhanbad town. The weddings with Suresh brought her to Hadi Cooli. Her parents had offered financial support to buy a piece of land and construct her own house in Dhanbad. Despite the support provided by his father-in-law that would have changed his life, he chose not to migrate. Suresh is quite comfortable and close to his childhood friends and relatives living in Hadi Cooli. His father, Jamal Hadi, passed away, but he stays with his mother, Dulali Devi. The most important was the affinity and close intimacy with their birthplace, peers, and siblings, as he grew up in his birthplace, the peers, and siblings with whom he grew up and lived for years. It is important to be at one's place of birth and achieve a sense of satisfaction rather than migrating to another town. Suresh reiterates that he would not send his sons far away. He would try to educate them. Two sons are in the primary school located nearby.

Hadis feel that they will only get sweeping, rikshaw pulling jobs in all places and not better work, even if they migrate. There will not be any savings while working in a big city. The vulnerability is seen as higher for migrants. Hadis do not face a crisis in getting a sweeper job in Chas and Bokaro. Hadis also feels that new companies and offices are increasing in the area, which would increase employment opportunities for them. Until almost 30 years back, the Hadi buried their dead; of late, many of the burials done have had a Muslim influence. The influence is seen in the history of funeral practices and in giving names to children as well.[2]The geographical proximity and social interaction among Somesh's parents were better, and later, they lessened. Shyamlal's study on Bhangis in Rajasthan also shows that in the past, Bhangis buried their dead. He finds that Bhangis' burials are gradually changing into cremation. Only young children and those who die due to smallpox are buried because Sitala Mata sends this disease.

Hadis, too, believe in Sitala Mata, which makes them worship her. They believe this may help them to avoid major diseases like smallpox, which had a high prevalence in the early 1980s. Nicholas (1981), in his paper 'The Goddess Sitala and Epidemic Smallpox in Bengal', wrote how Goddess *Sitala* has an all India presence except southern states. He further elaborates that Westerners tried to eradicate smallpox. The eradication of smallpox would help in the disappearance of worship of the Sitala goddess, if not the Goddess. According to Nicholas, Westerners had less idea to understand the goodness arriving in the body, and Indians ritualised the terrifying conditions during smallpox. He described how Sitala emerged in the medical texts in India in the sixteenth century, but there is less clarity on when exactly it became associated with smallpox. Her image portrayal in "A temple at Modhera in Gujarat shows her mounted on an ass, naked and winnowing fan in hand, another figure in the outer wall of the twelfth-century temple at Sejakpur in Saurashtra" (Nicholas 1981, pp. 29–30) gives the presence of goddess Sitala at all India level. Nicholas does find *Sitala* in his fieldwork area of Midnapur. He further writes the depiction of *Sitala* during the eleventh to twelfth century, where she is portrayed having a broom in one hand, riding an ass and one winnowing fan in the left.

The presence of the *Sitala* temple in Bauri Cooli shows the similarity of her presence in most parts of eastern India. Babagrahi Misra (1969), in his paper 'Sitala: Smallpox Goddess of India', describes Sitala as a disguised woman who went to the king of Chandraketu in Bengal and asked him to worship her. People in Bengal were not deities to Sitala, which made her consult a fever demon who brought 64 kinds of fever. He reaches on earth and puts this fever into the human body. All these depictions manifest the centuries-old belief about Sitala and her role in bringing cures to the human body. People in Hadi Cooli believed *Sitala* was a grace to them

---

[2] Jamal Hadi, Gulam Hadi, and Omar Hadi are a few Hadis' names of the houses who had closer interaction a few decades earlier. Hadis women would also stand outside to the mosque to get *Dua* and *Phook* from the Imam but this has declined now.

and not a disguise. Due to this, people do not want to be cured with medicine but rather perform rituals and pray at her temple.

The dead body was buried in a faraway area at a distance of 2 km from Hadi Cooli and not cremated. Somesh and Gulam said, "It was accepted as a better way to perform the last rites of a person following Muslim rituals. We would dig the trench ourselves and put the body inside. Our forefathers were closely associated and had a good relationship with Mohammad Tukan's ancestors in Ansari Mohalla, who was the Zamindar at that time. Gradually, we started going to check the cremation (dahaan) area on the Garga River side. However, I do not know much about this in detail."

The death of any member adversely affects the head of the household. The funeral ceremony is an occasion of burden due to the expenditure involved. The mourning and grief are two-fold; firstly, there is a social responsibility to organise a feast for the community members after the funeral ceremony. Secondly, this ritual sometimes forces households to borrow money. The debt is taken from moneylenders based in Sahu Bazar. Baniyas, a few Sikhs, and Marwaris from Rajasthan primarily own the town's market. Nevertheless, Baniyas, locally known as Sahuji, have a good grip on Chas town's retail grocery and wholesale markets.

The dead body is placed on firewood, set afire, and burned to ashes. There are two funeral sites in the town. One is located on the bank of the Garga River. One of the oldest rivers in the Chas-Bokaro region, the river is also considered ritually pure and significant for Hindus. Hence, a large cemented cremation area is constructed on both sides of the river with concrete near the Check-post by the Chas Municipal Corporation. Hadis find it convenient to cremate here. It is located away from Hadi Cooli and is around 3 km away. Another site is *Joria*, which is located in the vicinity of HadiCooli. Over the last 10 years, the Joria has turned into a stream carrying all the sewage water going down all year round. The water is very filthy and polluted due to industrial and household drainage. However, during monsoon, the rainwater cleans *Joria* to an extent. Land encroachment on both sides of *Joria* is another reason for the declining water level in *Joria*. The agricultural and open *Gair Majurwa Zameen* (lands) is being encroached; this has also closed the route of water flowing towards the *Joria*. The stream is not advisable, but most Hadis use the water for several purposes. Cremation activities are not practised nowadays near Joria due to the unavailability of land and the residents' objections. Somesh once said that some of their elders were buried on the outskirts of Cheera Chas village. Cheera Chas was a village until a couple of years ago, but now it has become part of urban Chas.

The community members believe that it is not essential to cremate the dead body of a Hadi. Somesh Hadi narrates that one of his paternal uncles was buried in the way Muslims bury the dead.

2.15.4 Hadi Youth Association

Hadis is trying to organise under one platform, and they have formed an organisation called '*Jharkhand Hadi Yuva Samaj Manch*' (Forum for Organisation of Hadi Youth of Jharkhand). The organisation's purpose is to safeguard the interests of Hadis in the state and assert their rights under one common platform. However, the presence of the organisation has minimal visibility in routine activities. Sudhir Hadi, one of the organisation's key members, conducts meetings whenever he feels to meet. Health, education, and employment facilities are also important in the organisation's annual meetings, as articulated by Sudhir Hadi, the president. The issues taken up in this background provide some idea about the living conditions of the Hadis. The youth are conscious and aspire to have a better share of government schemes: education, employment, and health. These issues are discussed regularly in the scheduled meetings every Wednesday held at the community hall cum temple in Hadi Cooli. The meetings are restricted only to the Hadi youth.

The Hadis in Jharkhand are organised as caste associations; at times, there is a state-level annual meeting of Hadis. Similarly, another organisation, called 'Akhil Bhartiya Navyuvak Dal' (All India Youth Force), is run under A K Roy, a left political leader. The recent municipal elections in Jharkhand have shown a growing interest among the Hadis in reorganising themselves and creating a pressure group for lobbying. Such a move succeeded with the Shakaro caste in the Chas Municipality election 2010. All the Shakaro caste members from each family united and promised to help the candidate win by providing all support from their caste members. As a result, this was the only one whose members could win the election as a ward member of the municipality. The smaller number of Hadis makes others sideline them, and they cannot win in the ward election. However, in consultation with other elders, Ramesh Hadi decides who should receive the en masse Hadi vote. On the question of discrimination and occupational pollution, in a group discussion meeting with the Hadi youth, it was observed that they reject the notion of Hadis as untouchable and do not accept others' views that Hadis are polluted and untouchable. Such changes have prepared them to look beyond Hadi Cooli and spend their life outside the castes, even if limited.

The youth have leisure time during the day after they return from work. They can be seen spending time with friends in the main market area. Some would sit and chat at the corner of the Hadi Cooli, adjacent to the main road. Several of them were found to be playing cards or Ludo.[3] It is an indoor game with a maximum of four participants, with peers. During one of the discussions, it was recorded that, at present, there is a lack of space/ground where youth can play football. Football remains their all-time favourite sport, except for a few who also love cricket. However, most youth admitted that they watch cricket matches when India plays. They attend the football matches held in Telgadia village (4–5 km from Hadi Cooli). The surrounding villages of Telgadia organise a football tournament annually. Youth and children from Shakaro,

[3] Hadi Youth play Ludo, sitting at the road side. They also play cards on the road side among the peers.

Bauris, and Muslim localities play football and other sports on an open field next to Bauri and *Ansari Mohalla*. It is far from Hadi Cooli. Hadi's children are not encouraged to be involved in football play. The non-availability of any ground near Hadi Cooli, Hadis, was found to be secluded in the evening hours. Perhaps it is another reason that Hadi youth and children play games outside their Cooli and adjacent to the main roads. Many of them stand in groups and chat among themselves. Waking up early in the morning and searching for work is the daily and regular routine of many Hadi youth and adults who do not hold regular jobs. If they get some work, it provides them with pocket money for tea, snacks, cigarettes, alcohol, etc.

The children from both Bauri and Hadi Cooli have a common ground at the market to intermingle and establish social interactions. A few Hadi youths visit Bauri Cooli in the evening and have a good time at the tea stall with *Pakodi* (a kind of snack).

2.15.5 School and Educational Interest

A primary school is located on the outskirts of Hadi Cooli, near Bauri Cooli and Muslim Mohalla. The school has children from all three localities: Hadi, Bauri, and Muslim. There are five teachers at the school. It is a small building with four rooms and a large veranda. In front of the school, a deep well was sunk as a water source, which the Muslim Mohalla and Bauri Cooli used for drinking water. However, it has now dried up and is filled with garbage, leaving the school without a water source.

A good number of children are enrolled, but they rarely attend classes. During mid-day meals, the children return to school and enjoy the food. Teachers find it difficult to maintain regular attendance. One teacher stated, "Teaching children from lower castes is challenging. Their parents do not bother about the progress of their children. Most parents from Hadi Cooli do not visit school, and even when their children miss school for a month sometimes." The educational level of Hadis is relatively low, as shown in Table 2.5 in this section. A larger number of them are illiterate, both male and female. No single woman who reached class 11 is found for Jhomda Hadi, who completed graduation.

The lapses in regular schooling are another serious problem that has improved Hadi's performance. After primary schooling, only a few enrolled in the high school located 3 km east of Hadi Cooli, and they rarely attended. Apathy towards education is prevalent in Hadis's mindset, including among the youth. The youth is the son of

Table 2.5 Educational level of schooling among the Hadis by gender

Education Sex	Illiterate	Class I–IV	Class V–X	Class XI and above	Total
Male	72	40	23	4	139
Female	123	27	17	Nil	167
Total	195	67	60	4	306

a tea stall owner, who studied in a common school, Ram Rudra High School, and qualified for an officer's job (Probationary Officer) in the State Bank of India. In interactions with young children, they showed little interest in discussing education.

There is a strong feeling that no jobs are available to them; ultimately, they will have to return to their caste and traditional occupation. I asked Krishna and Abhishek, son of Anup Hadi, why they did not pursue schooling when their father completed the sixth class; both described it in their ways. Krishna immediately responded, "Even then, my father is doing the same work." In contrast, Abhishek provided another perspective. He said, "I know a few Bauri youth who studied until class 12 but never found any government job. They also tried to apply for recruitment in the Jharkhand Police but failed. The recruitment in the Jharkhand Police is tough due to the essential need for physical fitness. Even if we pass class ten, our physical fitness is not up to the mark for police recruitment. However, there is a preference for SCs, as my friend tells me, but how do we improve our fitness with a minimum meal daily?"

Hadi's youth can provide many reasons for not attending school. Indeed, there is a strong feeling that a sweeping job is always available for them, if not any other job. The attendance of children in government primary schools is high. In front of the school, a deep well was sunk as a water source, which the Muslim Mohalla and Bauri Cooli used for drinking water. However, it has now dried up and is filled with garbage, leaving the school without a source of water. The school teacher found that about ten Hadi children continued school for a year. There is a provision for free school dress, books, and stationery. However, some parents reported that students sometimes receive nothing except books. During interviews with teachers and other respondents, it emerged that even the mid-day meal is not provided correctly; it is not cooked properly and is of very poor quality. In class one, one child, Aman Kumar, says, "We have got bored eating khichdi every day. However, I eat because we do not have a choice at home." The small children lack aspiration. They do not associate with the existing school and only show interest in playing and roaming around the locality.

Educational backwardness is rampant among the Hadis. Almost every day, discussions on educational backwardness occurred in the field while I was there. The socio-economic condition does not allow them to spend on their children's education. Schooling is free for all SC children in government schools, and uniforms are provided. Several children received uniforms and books, but they are still unavailable to every Hadi student. Some parents appreciate this facility because the school provides uniforms. However, there is no scholarship for students in school. Some parents shared that their children do not remain in school during school hours but play outside. Some parents expressed concern about how they would know whether their child is in school, as they cannot attend during the day without leaving their housework and wage work. Housework is given a higher priority than children's schooling.

In contrast, the Bauri community has started sending their children to school, as they are aware of job opportunities in the government sector through reservation. However, Hadis remain unsure about such opportunities outside their traditional jobs.

Another government girls' high school is located 3 km from the check post area. However, none of the Hadi girls are enrolled in this high school. A strong perception among Hadis exists that degrees will neither bring prosperity nor feed them. They must work daily without fail to earn a minimum amount. Food and clothing are essential items in the lives of every Hadi. In the whole of Hadi Cooli, only one, Jhomda Hadi, is a graduate, but he is perceived as unsuccessful. He tried to find a good job, but a simple graduation degree did not help him. He was unwilling to revert to the same sweeping and scavenging job. It was tough for Jhomda to survive in an environment where education held little significance in earning a living. One Bangla (Bengali) medium middle school is located near Joria, which gives Bauri Cooli an edge due to its location. There are no success stories in the lives of the Hadi community. Most parents do not aspire to encourage and educate their children, as it is believed that after years of education, there are still no job opportunities. Their only hope rests on their children if they can overcome the deprivation and attain higher education. A strong belief among most Hadis, except for a few, is that educating a child will not bear fruit for the household. Those parents working in BSL, MADA, and a few others have a better understanding of the future of education. I recall Goli Hadi sharing his grief. He was married at 13 when he was a child and is now 52 years old. He said, "I had no idea about marriage; even my wife was very young (7 years only). However, our parents decided and did our Bartu (engagement) and after a few months, Viyah (marriage). But my wife came to my house a few years after the wedding when she was around 13, as I recall. My father told me that I should work and earn for the household. However, I completed an intermediate in commerce (XII). I would have pursued higher education and found a better job than sweeping, but marriage did not allow it, and I started working at an early age. My brother was also married early. But ironically, my son did not finish primary school and left despite my strong objections. The environment in Hadi Cooli is not conducive, as most children have little social interaction and friends outside their caste. I only got a job at MADA because they could not eliminate my application. There were other Hadis but without a matriculation degree. Due to a lack of knowledge and qualifications, I could not attempt a job at BSL during my youth. Now the competition is tough, and many other SCs compete in the sweepers' category. Goli, Pintu, and Ganesh are a few parents who want to educate their children at an English medium school in the Chas main market area, half a kilometre from Hadi Cooli. The school charges a fee of Rs. 100 per month, and for three children from one household, the fee is reduced to 200 rupees per month. Goli thinks the school is better than a government school; children are learning English. Their jobs in MADA and Pintu's in-laws from Dhanbad have influenced the decision to send the children to the English medium school. It is hoped that they will study well because of the good school.

2.15.6 Children in the ICDS Centre

One Integrated Child Development Services (ICDS) centre at Bauri Cooli serves children from all Scheduled Caste (SC) groups. The presence of an ICDS centre in the locality of Hadi Cooli is viewed positively, functioning as a link between households and the school. Here, children receive some learning and undergo early socialisation. Two women, including one cook and a helper (Anganwadi helper and Anganwadi Woman Worker-AWWS), are employed at the ICDS centre. The centre has 50 children enrolled, but they typically stay for about an hour while food is prepared and served. The children arrive at 11 o'clock in the morning, and when the food is ready, around 1 o'clock, they have their meal and leave. Hadi's children have the lowest attendance and do not stay long. Occasionally, they miss the meal because they arrive very late, around one o'clock or after. Hadi's children are unaware of why the Bauri or Rajwar children treat them unfairly or differently. Although there is no overt discrimination at the centre, children generally do not interact with one another outside of it. The caretaker mentioned that children from various caste groups mostly play in their own streets, feeling more comfortable within their own groups. She denies any caste-based discrimination against the children. The following picture illustrates the potential for children to learn to coexist, but this experience is limited to the centre.

In Photograph 2.4, taken during the fieldwork in May 2010, children anxiously await their turn to receive food from the centre's caretaker. Six of them, whose faces are visible, belong to Hadi Cooli. The small kids play at the centre until the food is served. During the first few visits, the kids would stand up after they realised my presence. However, when this photograph was taken, they did not bother and instead waited for their turn to have Khichdi (meal) on their plates first.

2.15.7 Changes in Rituals and Festivals

Rituals are similar to Hindu rituals but are not participatory, as other caste groups are not included. There is increasing interaction, but inter-caste relations do not exist at a social level, except for a few individual interactions. Of course, there are no clashes, competition, or hassles between them, but all the caste groups live and perform their daily rituals separately. Some Hadi members engage in daily religious activities. A few young members were found tying threads at sacred spaces on the wall shelf during prayers, referring to them as Ishtdeva. They pray for their betterment, keep the deity happy, and seek blessings. The nearest Shiva temple exemplifies the gradually increasing inter-caste closeness during festivals and Puja.

The adjacent temple has provided some space for social interaction among other castes, but it is minimal. This temple is situated between Ansari Mohalla and Hadi Cooli, touching the walls of Somesh Hadi's house. The Hadis constructed this Shiva

Photograph 2.4 ICDS centre, children waiting for mid-day meal. Source: Ziyauddin (2022)

temple. However, other caste members residing in the nearby market area lent financial support, including the Sahu caste members, Bengali upper castes, and some Bauris. It was observed that individuals from other castes visit the Shiva temple of Hadi Cooli every day. Nevertheless, on the day of Shiv Ratri, many people turn to the deity in the temple. Women visit the temple when they are free from household chores. One priest is responsible for cleaning and conducting rituals during Puja. Bijla Devi, a 30-year-old woman, says, "I do not get time to go to Shiv mandir, but I do pray to Mansa Devi in the morning. She is a supportive Goddess in our household. I perform a sacrifice during the Mansa Puja annually."

Photograph 2.5 of the Shiva temple, later in this section, shows minimal movement of the people. I took the picture at noon in January 2010. However, Hadi's kids play in front of the temple in the evenings during the summer season, a scene captured in the photograph. Apart from this area, there is no open space available for Hadi's children to play and spend time during the evening. With trees surrounding it, the temple also provides shade in summer. The four trees near the temple are ancient and have been cared for by the priest. Otherwise, such old trees would not be found. They are the oldest trees in the locality of Chas, as most have been chopped down to expand new housing areas. The temple priest lives in one of the small rooms along a veranda just in front of the temple.

Hadis have their festivals and rituals. In a small but significant way, the festivals and rituals of the migrants and other caste groups have influenced Hadis and their ceremonies. However, most festivals are commonly celebrated as they are also found among other Hindu caste groups. Mansa Puja is a significant festival celebrated on a

Photograph 2.5 Shiva temple in Hadi Cooli. Source: Ziyauddin (2022)

large scale among the Hadis. The most important aspect of this festival is the sacrifice offered to the Goddess Mansa, as a small animal or bird. Hadis believe her to be the protector of their life and lifeworld. The lifeworld of Hadis refers to the reality of the conditions and circumstances in which they live (Krauss 2015). There is a substantial dependency on the goddess, who is believed to help them in dire need. Everyday life presents various challenges, but these are not uncommon. Women often find it most difficult due to various exigencies and seek solace in prayer to their deity. The daily tasks of cooking, working, and other natural and habitual activities are essential components of the daily routine Hadis engage in. Daily life (Felski 2000) may be mundane and routine, reflecting how Hadis live.

The Mansa Puja festival is celebrated during the Asaaddh month of the Hindu calendar, typically between June and July. A new festival, gaining traction among native communities, is Chhat Puja. Chhat Puja is regarded as the most important festival for the people of Bihar and many in Jharkhand. However, the Hadis show less interest in Chhat Puja. Malti Devi describes Ramesh Hadi's wife, saying, "Chhat was not a festival for Hadis and even the neighbourhood Bauris. Nevertheless, many of us have liked to perform Chhati Puja over the last decade, and if not celebrated, we wish to participate. The migrants settled in Chas town place greater importance on it, and the entire Ganga Riverside is decorated during Chhat Puja. The majority of the residents in Chas town are Biharis. Still, it is not popular among Hadis in the area. We visit Garga's side and enjoy observing the festivities in large numbers."

During the months of Bhadra-Ashwin and Kartik, major festivals such as Janmashtami, Durga, Laxmi, Kali, and Kartik Puja are celebrated. Another significant aspect is that during Mansa and Laxmi Puja, women play an important role, as these are considered women's festivals. They pray for the welfare of their families and children. The festival involves fasting during the day and offering prayers to the deity in the evening at sunset. In the case of Durga and Kali Puja, women's roles are confined to the household arrangements for the Puja, while men offer prayers to the deity. During Mansa Puja, animal sacrifice is essential; if a goat cannot be sacrificed, at least a chicken or a duck is offered, as it is believed that Mansa Puja is incomplete without such a sacrifice. This depends on the financial capacity of the household. Families sometimes incur debt to purchase the animals and items required for the Puja. There is no temple of Mansa in Hadi Cooli, but they can visit other nearby places, such as the Bye Pass Road.

Life in Hadi Cooli is relatively different in several respects compared to the above. Hadis must follow their work, which demands different working hours than most other jobs. Community solidarity is evident, especially when elders pass on any information. New visitors will be asked about the purpose of visiting Hadi Cooli. Young children tend to spend time alone, except in the company of school-going students and friends outside Hadi Cooli.

The lives of Hadis are changing, but there is limited social mobility. Bhangis in Rajasthan have begun adopting the customs and traditions of upper castes, demonstrating the process of Sanskritization, as Shyamlal (1984) mentions. There is a diversity in the occupations of Bhangis in Rajasthan, but Hadis exhibit slight variation, with only two households engaging in different jobs; one person operates a grocery shop within his house, while another works as office staff at Tata Indicom Retail shops in the private sector. In 'The Bhangis', Shyamlal (1984) found they were engaged in nine different occupations. The study is 30 years old, when private sector employment opportunities were limited in a town like Jodhpur. However, the growth of the industrial and urban sectors could have increased opportunities for Bhangis and has also benefited Hadis in Chas. Despite this, Hadis in Chas have retained their traditional occupation of sweeping rather than diversifying into other jobs.

Therefore, this chapter summarises how the Hadi community's social, historical, and occupational trajectories as a Dalit group in Chas town illustrate the intriguing and ongoing intertwining of caste-based stigma, structural marginality, and economic disenfranchisement. This occurs despite planned urbanisation and legal prohibitions against discriminatory practices. The data indicate a process of incremental change in the township areas through a limited mechanised sanitation system, employing labour on a contractual basis within the municipal corporation, and shifts in public interaction that have altered specific modalities of exclusion. The narratives and informed interview-based data present a perennial association of the Hadis' identity, deeply rooted in the genesis of labour division based on birth and heredity. This regime has survived for centuries among the Hadis. The experiences of the Hadis, as one of the most deprived Dalit groups in Chas town, highlight the inconspicuousness of subaltern Dalit sub-castes within the state's enumerative frameworks and urban policy discourses, revealing the limitations of institutional recognition and welfare interventions. Therefore, this chapter calls for a reorientation of urban sociology and development policy to engage with caste as a structuring force critically and to centre the voices and struggles of historically oppressed communities, such as the Hadis, in both scholarship and governance.

References

Atluri, Tara. 2011. You Marxist I Clean Toilet: Racism, Labor, and the Bathroom Attendant. *Frame: A Journal of Visual and Material Culture.* Spring (1): 69-95.

Census of India. 2001. Basic Data Sheet: District Bokaro- Jharkhand. (13), accessed 19 April, 2015. http://censusindia.gov.in/Dist_File/datasheet-2013.pdf.

Census of India, 2011. *District Census Handbook.* Jharkhand Series-21. Part XII-B. New Delhi: Registrar General of India.

Chas Master Plan, Vision 2027. Chas Municipal Corporation. accessed 10 August 2015. http://chasmunicipalcorporation.com/AboutCMC.aspx.

Chas Municipal Corporation. 2027. GIS-based master plan for Chas. *Urban Development & Housing Department*, Government of Jharkhand.

District Statistical Handbook, Bokaro. (2014). *Government of Jharkhand.* bokaro.nic.inindiastatdistricts.com

District Statistical Handbook, Purulia. (2014). *Department of Planning and Statistics*, Government of West Bengal. Wikipediapurulia.gov.in

Felski, Rita. 2000. The invention of Everyday life. *Cool Moves-A Journal of Culture Theory Practice.* 39 (9): 15-31.

Freeman, James. 2004. *Collecting the Life Histories of an Indian Untouchable.* In Vinay Kumar Srivastava (ed.) Methodology and Fieldwork. New Delhi: Oxford University Press.

Goswami, B. N., and Xavier, P. (2005). Dynamics of "internal" interannual variability of the Indian summer monsoon in a GCM. *Journal of Geophysical Research*, 110. [Note: Volume/issue/pages not available—if you have them, please share for completion.] ResearchGateNatureACP

Harshe, Rajan. 2014. Review-A History of Prejudice: Race, Caste, and Difference in India and the United States, accessed on 7 Aug, 2016. www.cambridgeindia.org/reviews/Seminar

Jharkhand Municipal Act. 2011. Jharkhand Act 07 of 2012. Ranchi: *Jharkhand Gazzette.*

Jhosua Project. 2014. *Hadi in India*, accessed on 14 Jan, 2015. https://joshuaproject.net/people_groups/12027/IN.

Krauss, Bjorn. 2015. 'The Life We Live and the Life We Experience: Introducing the Epistemological Difference between 'Lifeworld' (Lebenswelt) and 'Life Conditions' (Lebenslage). Social Work and Society. 13 (2).

Lahiri-Dutt, Kuntala., Radhika Krishnan, and Nesar Ahmed. 2012. Land Acquisition and Dispossession Private Coal Companies in Jharkhand. *Economic and Political Weekly.* XLVLI (6): 39-45.

Lok Sabha Secretariat Parliament Library and Reference, Research, Documentation and Information Service (Larrdis). 2013. *Manual Scavengers: Welfare and Rehabilitation-Reference Note for Members of Parliament*, No. (18): Rn/Ref./August.

Michael, S M. 2007. *Dalits in Modern India: Vision and Values.* New Delhi: Sage Publications.

Misra, Babagrahi. 1969. Sitala: The Small Pox Goddess of India. *Asian Folklore Studies.* 28 (2): 133-142.

Nayak, Pulin. 1995. *Social Exclusion in India.* Geneva: International Institute of Labour Studies.

Nicholas, Ralph W. 1981. The Goddess Sitala and Epidemic Smallpox in Bengal. *Journal of Asian Studies.* 41 (1): 21-44.

Pandey, Gyanendra. 2013. *A History of Prejudice: Race, Caste, and Difference in India and the United States.* New Delhi: Cambridge University Press.

Patel, Tulsi. 2008. Stigma Goes Backstage: Reservation in Jobs and Education. *Sociological Bulletin.* 57 (1): 97-114.

Pathak, Bindeshwar. 1991. *Road to Freedom: A Sociological Study on the Abolition of Scavenging in India.* Delhi: Motilal Banarsidas Publishers Pvt. Ltd.

Ramaswamy, Geeta. 2005. *India Stinking: Manual Scavenging in Andhra Pradesh and their Work.* Chennai: Navayana.

Russel, R. V. and Rai Bahadur Hira Lal. 1916. *The Tribes and Castes of the Central Provinces of India.* Vol. 4. London: Mcmillan.

Sachchidananda. 2002. *People at the Bottom: A Portrait of Scavengers.* New Delhi: Concept Publications.

Sengupta, Nirmal. 1980. *Destitutes and Development: A Study of the Bauri Community in the Bokaro Region.* New Delhi: Concept Publishing Company.

Shah, A M. 1998. *The Family in India: Critical Essays.* New Delhi: Orient Longman.

Shah, A. M. 1996. Is the Joint Family Disintegrating. *Economic and Political Weekly.* 31 (9): 537-542.

Shyamlal. 1981. The Occupational Mobility among Urban Bhangis. *Eastern Anthropolgist.* 11 (1): 11-20.

Shyamlal. 1984. *The Bhangis in Transition.* New Delhi: Inter India Publications.

Shyamlal. 1992. *The Bhangi: A Sweeper Caste Its' Socio-economic Portraits with Special Reference to Jodhpur City.* Bombay: Popular Prakashan.

Singh, Kumar Singh. 1995. *People of India: Daman and Diu.* Anthropological Survey of India. pp. 74-75. New Delhi: Popular Prakashan Private Limited, XIX.

Srinivas, M. N. 1942. *Marriage and Family in Mysore.* Bombay: New Book Company.

Srinivas, M. N. 1952. Religion and Society among the Coorgs of South India. Oxford: Clanderon Press.

Srinivas, M. N. 2003. An Obituary on Caste as a System. *Economic and Political Weekly.* 38 (5): 455-459.

Srivastav, B N. 1997. *Manual Scavenging in India: A Disgrace to the Country.* New Delhi: Concept Publishing Company.Srivastava, Vinay Kumar. (ed.) 2004. *Methodology and Fieldwork.* New Delhi: Oxford University Press.

Thurston, E. D. (1909). *Castes and tribes of Southern India* (Vols. 7). Government Press. JSTOR

Venkatesan, V. 2005. A Case for Human Dignity. *Frontline.* 22 (12). June-4-17.

World Health Organization. 2009. *Self-care in the Context of Primary Health Care: Report of the Regional Consultation Bangkok.* Thailand 7–9, January 1-2.

Ziyauddin, K.M. 2022. Situating Hadis' Occupation and Caste. CASTE: A Global Journal on Social Exclusion 3 (2), 263-284

Chapter 3
Perceiving Illness Experiences

3.1 Introduction: Viewing Perception

Perception is understood as seeing and experiencing something through individual senses or mind. Sensing objects, problems, physical movements, people's experiences, etc., helps interpret meaning according to individuals' understanding and cognitive capacity. How does an individual notice and understand that things that are not noticeable to another person vary from one person to another? When a person translates his or her sensory impressions consistently and gives meaning to the world he or she lives in, that is the perception of the person. The thesis takes the help of the concept of perception in order to understand perceived notions about illness and health and try to understand in the context of Hadi Caste in Chas town, Bokaro district. While interpreting the information to explain and understand the social situations and environments, the person uses the method of organising, identifying, and interpreting her/his sensory information. In a way, there is an involvement of the mind and thoughts that, mind and thoughts shape perception, like learning, memory, expectation, and attention (Gregory 1987).

In this process, perception differs from person to person, individual to another, but one or several people may perceive one situation differently or similarly. An important dimension shows that perception significantly differs from one individual to another but also finds certain commonalities about the outer world.

People from Hadi Cooli perceive the process of illness experiences differently from others. Hadis, in perceiving their world, begin constructing their perception while learning and doing everyday work. The Hadis, as individuals and as persons belonging to a low caste, infer and develop their own perceptions about several matters. The experiences during the social interaction and seeking health care services in clinics and hospitals are a few instances when a Hadi develops some understanding about their work, social interaction, and episodes of illnesses and gives meaning to them. By giving meaning to the information and experiences of a person, there is a possibility that perception differs from one person to another according to their

K. M. Ziyauddin, *Occupational Caste Groups' Illness and Health in India*, People, Cultures and Societies: Exploring and Documenting Diversities,
https://doi.org/10.1007/978-981-96-8967-5_3

information and interaction in society. For example, Hadis have their perceptions about the health care system, available hospitals, doctors, health personnel, and other health care providers whom they visit in hours of need for health services. Their perceptions are built in a process.

Therefore, the construction of perception is also a process of experiences in society resulting in the formation of "Illness perceptions which are organised cognitive representations or beliefs that patients have about their illness (Petrie et al. 2007, pp. 2–3).

There is another dimension in which perceptions determine health-seeking behaviour in various forms, namely illness, seeking treatment and finally, in which perceptions determine the health-seeking seeking perceptions determine the health-seeking perceptions in various forms, namely illness, seeking treatment and recovery from an illness. They make out various elements that include naming the illness, displaying ill health's indicators and faith or understanding about the reasons for illness. It is also significant to observe that Hadis has some understanding of how long an illness will last and that it is drawn upon their individual history or of their past illness. One's health condition or another member's illness is discussed at home. An agreeable level is reached about whether or not the treatment is to be sought according to the perception regarding the nature of the illness. As the detailed description of the illness experiences is provided later in this chapter, one notices that "Patients' beliefs about their condition are often at variance from those who are treating them" (Keith et al. 2021, pp. 2–3).

Further, while studying illness experiences, it is important to note that usually noted that most doctors and health care providers are unaware of the patient's understanding of their illness condition. It is considered unimportant to inquire about the patient's own ideas during clinical consultations. It is a different concern that patients' perceptions may significantly vary from one person to another. People carry and view different perceptions of common illnesses or ailments and diseases (Petrie et al. 2007). Hadis, in the present research, have shown various layers of perception to a common illness and the same health care centres.

Perceptions of health and illness are affected by various processes, and they, in turn, become factors of influence on individuals or communities. In this chapter, I will look into Hadis' various experiences and attempt to understand and analyse them.

The way Hadis perceive themselves is a crucial factor that influences their view of others' perception of them. The interaction of Hadis with various health institutions plays a pivotal role in constructing their perceptions. For instance, encounters at the reception of a clinic or a hospital, either government or private, the use of words during the consultation process with an attendant outside the doctor's chamber, and finally, patients' face-to-face interaction and narration of being ill to a doctor in his/her chamber, are certain levels of interactions that in turn affect the mindset of a Hadi patient. Hadis is found to carry dual identities: first, an identity of a patient and second, an identity of a patient and second, a socio-cultural identity as a Hadi who performs menial and polluted jobs. Perceptions of illness and health are deeply rooted in the customs and traditions of a community (Banerji 1981a, b, c). For example, Hadis perceives staying active as quite essential. An inactive and unwell body is perceived to be affected by an inability to attend to everyday tasks. Illness is cured at *davaee dukaan*, and the clinic of Doctor Vimal Singh is perceived among most Hadis in Hadi Cooli rather than delaying. An inherently unequal relationship between a doctor and a patient exists in consultation and treatment-seeking. The idea of Banerji (1989) in his paper on 'Rural Social Transformation and Changes in Health Behaviour' about the health behaviour of the people is relevant. He studied 19 villages in eight states. The study lasted for 15 years, starting in 1972. Banerji has discussed a wide range of issues, including health practices, the health-seeking behaviour of the people, and the importance of the availability and accessibility of public health systems to the poor population in his study area. He found that people gradually started becoming dependent on allopathic medicine with the presence of public health institutions. Lack of sufficient provision of health care treatment in the form of public health support also paved the way for large numbers for RMPs, usually for allopathic medicine. People gradually started becoming dependent on RMP treatment that was available to poor patients.

Following Kleinman, understand the distinction between illness and disease. Arthur Kleinman wrote,

> Illness is inherently an experience about human sufferings and symptoms. Illness is referred to as how a sick person and the members of the family or wider social network perceive, live with and respond to those symptoms and disability. Illness is the lived experience of monitoring bodily processes. The patients' communication about the entire process felt in the body is an integral part of the itinerary of the illness." (Kleinman 1988, p. 4)

This process of feeling certain abnormalities in the body does not get proper space at the trained doctors' chamber, unlike an RMP and, namely, traditional healers. Kleinman further illustrates that illness is also important when a common man illustrates the feelings and symptoms that are understood by the layperson's expressions in the social group they live in. Further, while talking about illness, they also use their judgement about how to cope and manage problems of illness in life and everyday activities.

Kleinman (1988) sees from this point when people think of seeking treatment that shows the illness behaviour. Illness for Hadis then leads to initiation for treatment, managing food and activities, eating better food than the normal days, monitoring bodily symptoms, and then going to a medical shop to buy medicines or visit a doctor to get medication. The illness behaviour of Hadis shows significance when they buy *over-the-counter medicine* (OTCM), and sometimes they are found seeking traditional/practitioners' services. At this juncture, illness complaints are noticed at the household level, and the same is brought to a practitioner. When a Hadi visits an RMP and seeks treatment, it reflects the illness's behaviour, and in turn, the practitioner provides a diagnosis and gives some name to the illness. It is likely to inform about the possible disease in the body. There is also some possibility that symptoms of illness are different and vary from the disease found in the biological body. The diagnosis pattern and nomenclature of disease is based on a bio-medical system of practice, not necessarily what is felt and suffered by a patient (Kleinman

1988; Kumar and Mahapatro 2010). Kosa and Robertson (1969) view the social implications related to a disease as playing a crucial role in deciding the illness behaviour of the patients.

However, the traditional healing methods have continued to be important ways of giving solace *Jhad-Phook* by a *Baba.* Several of the patients sought help from the Baba, Tantrik and also home remedies for minor illnesses, e.g. body aches, indigestion, cold and cough, headache, mild temperature, etc., showing their faith in traditional healing systems. Apart from this, the visibility of modern medicine is dominant in providing cures for illnesses to Hadis. The quick healing methods prompt them to depend upon medicine over the counter and at the clinics of Doctor Vimal Singh and Doctor Ganesh. Therefore, a gradual dependency on allopathic drugs has gained popularity over the last 20 years.

This gradual change in health behaviour reflects the perceptions of the Hadis. The change in health behaviour is reflected later in the narration, which illustrates Hadis' perceptions of seeking treatment. Shailender Hadi effortlessly narrated his illness at the RMP's clinic. He said, "*Petva mein badi kankani hovo hai, darad hai kii aare kutchu hai tonhe dekh. Garam paani peeliye ke kam hotae. Ab to konho aare ilaaz nae hai, pehle konho upaaye hovo hathlo budha logan ke kintu ab to yehe davae kaam kartae.* (I have too much of discomfort in my stomach. It is a pain or something else I do not understand. I also had lukewarm water to reduce the discomfort. Earlier, our elders had some traditional methods to subside problems but now it is only this medicine (allopathic) that can help)."

The belief and traditional practices to deal and cope with illnesses reflect the health care services sought and utilised among Hadis that how to manage health breakdown, health care practices, and preferences to health care treatment. Hadis were not exposed to modern medicine till the 80s. Some elders said their Budha-Purna knew a lot about herbs that could cure certain illnesses. For instance, Bawaseer (piles) was fully cured by some herbs to be eaten by the person suffering that still has no permanent treatment in modern medicines. Shailender said it is pretty common in our job to have injury, abrasion, or wounds that could deteriorate at times. Some grass is available around Cheera Chas fields that could cure any wounds. But now we have no idea about those grass and treatment. Mostly, we consult Doctor Vimal Singh to see if the cuts hinder our work."

Till the early 1980s, a large number of cases of smallpox were reported among Hadis, as told by Shailender Hadi. He said, "It was quite disturbing to almost everyone in Hadi Cooli to deal smallpox with the threat of smallpox." I was told that even Bauris were falling ill with smallpox. The community members felt that constructing the Sitala Mata temple and offering prayers would reduce and manage the threat of smallpox. This assumption of relying upon the temple was among Hadis and other castes, including Bauris. It is important to explore the myriad forms of illness perceptions among them.

Hadis think that their occupation acts as a barrier and social constraint during the medical treatment of an illness. Care providers identify them as Hadis who are polluted as they do sweep jobs in Chas. When Hadis reveal their caste identity to the medical and paramedical (or health care) staff and doctors during treatment, they experience different treatment. They are often given less time for interaction with the care providers and have to wait for a longer time (Acharya 2010a, b; Acharya et al. 2015; Patro and Acharya 2016). People who have lived in the town for many years or were born in Chas town are familiar with Hadi's occupational background. They are also easily recognisable.

Kabir Hadi works in Chas Nagarpalika. His wife, Sonam, hails from Hazaribagh. She found a different life in Hadi Cooli. Sonam shares that there was very little interaction by the Hadis outside of their caste, which was different from the situation in her maternal home place. She had friends among other castes as well, a fact that was rarely found among most Hadis in Chas, except for a few young children who do interact with their age mates outside Hadi Cooli.

Sonam shared the experiences of Neelam Devi, wife of Kalo Hadi, who works as a maid in an upper-caste house. Neelam does sweeping and wiping of the entire house, but she is not allowed to enter into the kitchen area even for sweeping. Sonam finds several occasions that remind her that she belongs to the Hadi caste, not just as a worker or a woman. This data provides some reflections that there is a certain level of social distance maintained when a Hadi interact with upper castes for different purposes.

Another story about Kamal Hadi shows that discrimination exists towards Hadis. Kamal Hadi lives with his mother. Both of them narrated various problems while seeking treatment. Jhomda Devi, Kamal's mother, joined the discussion and said, "We are living in a desperate condition because a lot of much amount was spent on the health expenses of my husband, Sipeta, during his illness. We took him to several doctors in Chas and Bokaro when the illness was not recovering. He mostly had a dry cough, and this would continue for several minutes." Further she says, "*Kahiya ye bimaari bhale nae jano hiye hamni. Kutch baras to colliery mein kaam karal kae okre baad to hamni aa galiye Chas mein. Colliery mein dher garda-dhuaan hai. kaar ho jeebin saaf kapda pindhbin to. Kahan ki bhae ja hau hame bole parb!Ab hamar andar kutchu bimaari ghoosal hau to ham ki janbae.*" When did he get infected with this disease? We did not know. He worked for a few years in the mining area in Dhanbad, but later, we returned to Chas. There is too much dust and air pollution in the Colliery region. Your face and dress would get darkened there. I can't say what happens to people. If there is any disease in my body, how can I get to know?"

Jhomda Devi continued, "You will visit a doctor when there is a major illness; otherwise, why so? Sipeta was a man of courage. He would not listen and bother whether it was a cough or something else. We get several minor illnesses, but we generally think we should avoid consulting a doctor. Whenever you go to the doctor, it will cost 100–50 rupees, one has to spend unnecessarily. This was one reason that Sipeta initially avoided visiting a doctor. Even my mother would tell us to visit Dhanbad; she wanted him to consult at Sarai Dhela Hospital, a bigger hospital. However, he never listened to anyone." Kamal said, "I saw my father in his last days before he died due to TB. He had the courage to overcome the illness, but God had written something else. I had less time to spend with my father, and we could do little to save him." A grain of fatalism and helplessness is visible in Biru's summary.

This narrative shows aspects of illness perception. Besides other factors like lack of paying capacity or fear of expenses, ignoring and delaying the illness is affecting perceptions of illness and health care among Hadis. Sipeta was perceived to have unavoidable conditions and reported to the doctors about chronic cough, but it had already taken a serious stage.

The ethnographic data from the field provides sociological facts of social stigma and discrimination based on specific diseases. The negative social consequences doubled for those with certain diseases in the community and society. When Anulup's father was suffering from Tuberculosis (TB), it brought misery to their lives. Some community members would not visit his house. People would avoid any social interaction with Anulup due to the fear that TB will pass on to them. Anulup narrates, "The soul of my father must be feeling at peace, continuous cough throughout the day and night would almost kill him several times in a day, my mother used to cry at times. Who would wish that it is better to die than live in such chronic conditions? Why to live a life like this, my father would utter sometimes." Anulup's father, Jhagu Hadi, did get some treatment, but he was not given proper medicine on time.

3.2 Health Expenses

Financial conditions have a much deeper role to play in illness. It is related to the question of how a Hadi visualises her/his health condition. For instance, Hadi members live in perpetual financial crisis due to their low monthly income, which comes to an average of Rupees Rs. 3188 per month.[1] However, Chas Nagarpalika mostly pays rupees 4500 per month to those employed as sweepers. Out of a total of 51 households, 21 households in Hadi Cooli have a monthly income of less than rupees Rs. 5000, and these households form the largest middle-income group in Hadi Cooli. Those employed in BSL (two) earn higher; one earns Rs. 12,000 as a contractual employee, and another earns 19,000 per month. The remaining Hadis earn less than rupees Rs. 5000 per month. They are employed in different private offices (bank, hotel, office, etc.) that vary between three and five thousand monthly.

The average household earnings thus inform the fact that the expenses on food, health, clothes, and other essential needs leave them with no money in hand at the end of the every month. Hadis does not have the habit of saving for unavoidable reasons. The expenses of attending marriages, paying some time to the doctor, buying sweets or biscuits, and visiting a guest requiring arrangements to serve special food, snacks, etc., are a few of those needs that do not allow for savings.

One of the respondents, Susmita Devi, described recurring events in her past when she would save Rs. 100 and, at times, Rs. 500 out of her monthly income. Then, suddenly, her children would fall ill, and she would have to use all her savings to pay the doctor. She particularly recalls a day when she requested Doctor Vimal

[1] This average income is calculated from the monthly income of household members, calculated during the household census of Hadi Cooli in the year 2012.

Singh (known as Mandal) to give her the best medicine. She said, "During my malarial fever, my husband paid more than Rs. 1000 rupees as doctor's fees and for medicines. Injections and two bottles of saline were administered in the clinic, and an improvement was noticed. But after a few days, my son, Sunny, got a fever, and he was advised investigations, for which I paid more than Rs. 500 rupees." All her efforts to save were wiped out.

Recourse to treatment is limited, and free health care is also not desirable due to the distant location of government hospitals and the non-availability of medicine and doctors, Susmita said. Hence, most households pay extra money during their recurring illness. It is found that high dependency on local doctors has made them more vulnerable to paying extra amounts during emergencies. Until a patient's life is felt to be at risk, there is less likelihood that a doctor (RMP) would refer his patient to a better hospital. Doctors keep talking to patients and household members about the possible problems and causes of illness.

The longer the discussion, the better the trust and belief in the treatment of the doctor in the mind of Hadis. Susmita Devi could have visited trained physicians available in Chas at various localities, but she preferred consulting a nearby doctor. The preference appears more towards private health care than at government hospitals. It is noted that there is a gradual decline in the inclination to visit and access public health care services. Private health care services are in the form of RMPs, physicians qualified, trained clinics, and hospitals equipped with most investigations. The remedy to this problem is to avail treatment during illnesses at low cost and, at the same time, satisfactory treatment at Doctor Vimal Singh's clinic rather than in a government hospital. Employees in Bokaro Steel Plant have the provision of health care services at BGH. However, they mostly prefer consulting Doctor Vimal Singh to visit BGH due to its distant location from Hadi Cooli. BGH is located at a distance of around 12 km in Sector 9, Bokaro Steel City. Though medicine is free, the transport cost is higher than what Dr. Vimal Singh charges them.

The expenditure on illness is also a deterrent to the availability of treatment from other private doctors and hospitals. Hadis compares the treatment cost with a rough estimate and then decides on the treatment place. There is minimal increase in a Hadis' salary in comparison to Hadis employed in BSL. Most of their employers, like Chas Nagarpalika, Banks (like Axis, Indian Bank, State Bank of India, Bank of India), Post offices, hotels and lodges, nursing homes and hospitals, and other private offices do not increase salaries that can match the inflation. There has been a sharp increase in the cost of basic needs and food items in the last 5 years (2011). The inflation on main food staples such as rice in the wholesale market rose to 6.46% in September 2013 percent in August 2013. Prices of onions and vegetables grew 322.94% in the same month; in August 2013, the increase was 244.62%. The food prices rose due to Phailin destroying a flourishing crop in the eastern rice growing belt. This added to inflation due to high-growing plains being pummelled by floods in plain Bihar and Jharkhand choked India's inflation. This was in continuation of the high price of 2012. According to a report by Jute International, food inflation turned negative in the early year of 2012. Bihar and Jharkhand had high inflation rates of more than 9 per cent in 2013 (Kumar and Sinha 2014).

Mantu Hadi and his wife narrate their ordeal of high inflation in the last couple of years and the lack of state support provided to poor Hadis in the form of pensions, grocery provisions, etc. Mantu Hadi's wife, Monika Devi, a 45-year-old, has not been able to save even Rs. 20,000 to date. She grew up in the neighbouring district of Dhanbad in a small urban settlement called Lodna. She never dreamt that she would have no savings even after 25 years of marriage. Her parents hoped differently since Mantu had a temporary government job in MADA. But there are hardly any savings. Mantu lives with his mother and one unmarried son and his wife. His daughter Chandna is married to a family based in Dhanbad. His father, Shankar Hadi, is no more. His monthly salary is Rs. 11,000 rupees at a temporary job in a government office. The household's total income is Rs. 14,000 per month, including the monthly income of his son, Deepak. Deepak is employed as a sweeper in Chas Municipal Corporation (CMC). However, Mantu has still not been made a permanent employee as promised when he joined, and this has made his wife unhappy and dissatisfied. In a few years, her husband will reach retirement and will have to stop working. But he will be left with little money in hand. Their situation became tough after the separation of Bokaro as an independent district from Dhanbad in 1991 that is because MADA was in the jurisdiction of the Dhanbad district until 1991; it now looks after only the mines area within the Dhanbad district. Residents from Chas could not persuade the higher officials in MADA to absorb them as permanent employees. However, some employees who lived in Dhanbad were later absorbed as they could persuade the administration of MADA through the labour union. People like Mantu could not succeed in securing permanent jobs and continued as temporary employee jobs. Mantu recalls the need for money when he married off his daughter Chandni Devi. He said that they sold half of the jewellery given by Minika's parents to her at their daughter's marriage. Until now, neither of them has been able to buy any new jewellery. In brief, despite growing opportunities and a temporary government job, there is no positive change in their life. He has to support his daughter's household for her children's schooling. He is also repaying Rs. 100,000 that he borrowed at his daughter's marriage.

There has also been a gradual change in the food consumption patterns of the older generation, as narrated by Hadis. Now, Hadis do not eat carrion, nor are they forced to remove carcasses. In a group discussion with Ganesh, Ranjeet, Kaali, and a few others, it was noted that it is only less pocket money that restricts them from buying fresh chicken or mutton from the market. Whenever some festival or a guest visits their homes, there is a compulsion to serve good food and a non-vegetarian dish. *Murga-Gosh* (chicken meat) is common these days. During the late 2000s, the Hadis used to keep poultry at home, known as *desi murga*. According to them, it tasted better than the ones that they buy now. Ganesh says that chicken is served whenever we visit friends and relatives. But the desire to have mutton remains high. The families living in villages still raise poultry in a limited way.

Despite good awareness about nutritious food, low-income levels prevent them from cooking a complete meal, which includes rice, *dal*, and *tiyan* (vegetables). However, the most preferred cooked food is rice with vegetables, if available. A few mothers of infants and children reported wanting to provide their kids with healthier supplementary drinks like *Horlicks* and *Complan.* Little children also demand these costly supplementary nutrients, but parents cannot afford them.

The expenditure on food, clothes, education, and health for Hadis is relatively high. Almost the majority of households find difficult financial conditions. They find that expenses for food, clothing, education, and health amount to almost 90% of the household's income. Health is one area that can ruin the household's entire planning and eat into whatever small savings are made by the household. A few respondents reported that not less than Rs. 1000 is spent each time an illness requires treatment. Gita Devi narrates her ordeal, "I have to arrange and save money to marry off all three of my sons and a daughter. Whatever saving I had was spent on treating my daughter, Kajal. She would frequently fall ill with headache, nausea, vomiting, and an occasional bout of fever. I visited various doctors for her diagnosis, but nobody could cure her. Gradually, a doctor in the cooperative colony in Steel City treated her. By the grace of God, her suffering was reduced, and she is now 16 years old. In the last 10 years, I have not been able to save even Rs. 5000 rupees."

Hadis found it a worrisome task to manage monthly expenditures, as it demands a lot of calculations and proper management to run the household. It was also observed that priorities change monthly depending on the kind of need within the households. The small monthly earnings of Rs. 4000–5000 are not sufficient to accumulate savings in Hadi households. Buying rice, daal, and vegetables has become one of Hadis' most difficult tasks in the past 5 years. Kaali Hadi says, "Only a few Hadi households buy daal from his grocery shop. The rest go to the Hari Mandir market and try exploring for cheaper rates. Ultimately, many Hadis end up buying vegetables that are sold at lower rates, at the closing hours in the late evening but are inferior in quality. I also go to Hari Mandir and look for the cheapest vegetables. My wife adds extra spicy gravy to increase both the quantity and taste."

A new trend was noticed while discussing preferences for staple food. Youth are getting habituated to the staple food served in hotels such as Roti (made from wheat flour). Unlike the older generation, Hadis still prefers rice with curry and home-made pickles. For example, mango pickles are prepared during the season when they are cheap to buy from the Hari Mandir vegetable market. Roti, dal, curd, and vegetables in their meals are considered superior food among the new generation. They see these food items served in Chas hotels and compare them at their own houses. But most parents and heads of households are unable to serve such food. There is a day when vegetables are bought at relatively low prices during vegetable growing season. If there is good production and supply in the market, one can get cheaper vegetables during later evening hours. A few vendors want to sell off all vegetables and return home. Some women travel from villages to Chas market and try to sell their products by evening and return home early. Mitu Hadi says, "Mostly, I would buy vegetables in the late evening or around 8 o'clock when a few vendors just want to wind up the shop and sell their groceries at a lower price. The quality of the groceries and vegetables is good, in my understanding. Many Hadis follow this habit of buying vegetables. Even most grocery items are also available in the same market." There are variations noted in terms of choice of food.

Elders have no particular preferences except that meat is an evergreen choice if they can afford it. A sizeable number of eating joints in the form of *Dhabas* (road-side eateries) have come up due to the connectivity of the two National Highways and urban expansion, which could have influenced food consumption patterns. This tendency to adopt new food habits can be traced to those of upper-caste people and Muslims in nearby neighbourhoods who consume roti and sabzi for their meal at dinner.

Lunch and dinner are served properly. In the early morning hours, there is barely any time and cooked food is available to eat and go to work. A few Hadi youths get time to discuss while attending jobs in Mohallas and different residential colonies. For instance, Ganesh Hadi knows several families attend their cleaning and collect garbage at these houses. A few households would ask him if he could be given food and sweets when he visits in the morning. On such occasions, there are brief discussions between household members and Ganesh. This food-given-taken process happens with a few other Hadi workers while collecting garbage from houses. During festivals, there are occasions when a few households provide sweets and good food to them, but there is no certainty on all days.

Elders say that *Maad-Bhaat* (rice and its soup) is the staple food among Hadis as it is cheaper and cost-effective. The youth do not prefer but it depends on the earnings in the households. Ramesh Hadi added during one interview that *Madua* (Raagi or finger millet) was a popular food among poor households. It is grown and cropped during winter, but Hadis has no agricultural land. He says, "Madua is an extremely nutritious food. It is a staple food for the poor and many others. We did not have our own agricultural field, but I saw the cultivation of Madua (Ragi). We used to get Madua in exchange for some labour work. In our vicinity, Telgadia and Cheera Chas villagers would cultivate Madua as their main crop annually. However, today, you may not find Madua even in Chas market. You can eat roti from Madua flour or cook it in water while putting a little salt on it to taste. It was routine to eat Madua along with onion and green chilly in the morning and sustain almost for the entire day." It is a most healthy, and nutritious food for the poor.

A few Hadis said, "Our parents and elders used Madua for the post-partum mother either with milk or with salt and water. It is a healthy diet for children and lactating mothers. You will see in the near future people going back to Madua, but they will not find people growing up in Madua. Madua farming has vastly reduced in our region, nowhere in Chas." The change in staple food has also affected the health status of Hadis in general and women in particular due to its nutritious value. The field data shows that Madua is getting replaced with wheat rotis. The declining trend of Madua in Jharkhand is also due to the influence of wheat flour and its marketability and now madua has gradually reduced to only a minimal level of production and consumption,[2] so among Hadis.

[2] A report released by Ministry of Agriculture (Department of Agriculture and Cooperation) Directorate of Marketing and Inspection, Government of India titled '*Marketable Surplus and Post Harvest Losses of Ragi in India*' in 2002 suggests Madua to be an important crop and a healthy staple food for the poor farming communities. In India, the largest production is seen in the state

A few parents feel that they are not able to provide extra supplements in the form of fruits and meat to their children. In this context, even Doctor Vimal Singh reported "Children at Hadi Cooli are more vulnerable to viral diseases, and almost all of them contact it during peak summer seasons and just after monsoon. Malaria, typhoid and diarrhoea are the most common diseases troubling Hadi kids in particular." Vimal Singh says that those diseases also attack other people, but Hadis have a higher number of reports than others. This is due to the filthy and congested houses, drinking water is not pure, especially when there are leakages in the supply water, water gets contaminated, and Hadis use the same water. The supply water tap is close to an open *Nala* (drain). Many Hadis keep their utensils and pots outside the room or in an open area adjacent to their houses, making them infectious during monsoon. The open drain bypassing and the streets inside Hadi Cooli pose challenges and the possibility of higher infections.

3.3 Inability to Pay Doctors' Fees

The low-income level also restricts the amount Hadis can spend on getting illnesses cured by going to doctors. Reproductive health-related problems are least reported, as Doctor Vimal Singh informed. Women bring their children even for minor illnesses like stomach pain, uneasy feeling, fever, etc. and mention the symptoms of the illness. He said, "I hardly see any woman reporting for their own reproductive health-related problems. It seems they avoid their illness and think that not going to a doctor will help them minimise household expenses." This is the reason Hadi women develop much more weakness during pregnancy and post-delivery period than the Bauri women. This has been observed in various in-depth interviews with community members as well.

One respondent, Kabir Hadi (33), a sweeper at Chas Nagarpalika, highlights the pathetic condition in which Hadis live. He said, "If I do not go to work even for a day, my household dependents-children and wife will have to starve for that day. Every day, our presence is marked in the register. Do you know my job is private? Wages are calculated on an everyday basis. I have five children, all in the age group of 1–5 years. Who will pay the expenses for *suya-goli, as well as* doctors and medicines? Whether I am unwell or feel physical discomfort, I do not have a choice to be absent from work and lose my wages. I ignore the discomfort, but for the same reason. However, in case my children are ill, we make it a point to show them to the doctor and procure prescribed medicines on time."

In the above narration by Kabir Hadi, it is clear that parents' priority is to earn regularly and not lose wages due to absenteeism at the workplace. The sole earner of

---

of Karnataka where it is called Ragi. The report states that "retention for farm family consumption was the highest in Himachal Pradesh (87.46%) followed by Bihar (82.05) and Maharashtra (78.19)" whereas retention refers to the quantity of Raagi distributed into various sectors namely family consumption, consumption by permanent/temporary labour, estimated purchases, seed purpose and for payment by kind of the total Raagi produced.

the household has no choice but to constantly engage in some employment that pays. Kabir feels vulnerable due to his single earning status and the cost of treatment for illnesses that have to be paid. He also ignores his bodily discomfort and prefers to work. This is an attempt to minimise costs incurred for medication which is similar to many other Hadis. Until they are bedridden, most Hadis keep going to work. He says, "In case of fever which weakens the body, I am forced to rest for a day. One day's absence from work creates a money crunch in the household." He adds, "If I have one adult son to support household expenses, only then I can permit myself a day's rest; otherwise, it is very difficult and challenging to manage household expenditures, including health expenses."

The income situation among Hadis also varies depending on where they are employed. It simultaneously affects their perceptions towards seeking health care services, deciding the illness level, and preference over treatment sought. The following narrative gives a glimpse into this. Kabir compares his plight with those working at the Bokaro Steel Plant (BSL) and says that those working at the steel plant are in a position much better than the Hadis employed with Chas Nagarpalika. "BSL has provided many facilities to its employees, such as school, house, and health care. These can never match with Hadis working at Chas Nagarpalika." He then quotes a famous proverb to contrast the vast inequalities, "*Kahan Raja Bhoj aur Kahan Gangu Teli* (On one end is the comfort of a king and on the other is what a toiling human has.)" "*Nae pooch hau.*" (Please do not ask), he says.

Hadis who lived near the Steel Plant and those who lost land to the steel plant during displacement during the 1970s were given jobs. Three households who lost their land at Mahuar Basti (three Hadi households reside there) were also given employment at the steel plant. However, all of them work as sweepers in different branch offices and units of the plant. Two Hadis, Jago Hadi and Sudhir Hadi, living at the Mahuar Basti, were also interviewed. The village Mahuar Basti comes under the Jaridih Block Development Area, a block in the Bokaro district. While talking to Jago Hadi during an interview at Mahuar Basti, he said, "Basically, after getting the job in the Steel Plant, our life has improved, especially our children got a chance to study in its school at almost no cost. All my three sons have studied till matriculation and are now working at different offices of the plant but temporarily. All the children of my younger brother are studying in school and college. I also had staff quarters in Sector-9 of the Steel City colony but chose to live in my own house in my village. There are two small rooms in the BSL staff quarter, and we could not live properly there." His house in Mahuar Basti is larger than the allotted quarter, so he chose not to take the quarter.

Mahuar Basti falls under the Bokaro Steel City area but is on the outskirts of the township boundary. However, the village is not part of the residential colony known by Sectors 1–12. The three households have now expanded into five households. Jago Hadi is one of the most surviving seniors and has close ties with the families at Hadi Cooli in Chas town. Discussions with him helped me draw some comparisons between Hadis at Hadi Cooli and Mahuar Basti. It also brought out a probable understanding of the kind of health care available to BSL employees and Hadis not employed there. The life of a worker employed in a public sector unit is relatively better than that of a sweeper in private offices, Banks and Chas Nagarpalika. For example, Kabir Hadi finds it difficult to stay away from work, unlike those employed at BSL, who get leave during illnesses and receive free health care services at BGH. Whereas when Kabir Hadi falls ill, he avoids visiting a doctor and buying medicines.

Kabir said, "*Sardi khonki apnae baes bhae jeetae. Davaee khaae ke aare kamzor bhae jaebin hau. Doctarvan thheen jaebin toein kutchu to diyel hi partae na onkhni kae. Ek baat kahio, paache mahina hamar galva ekdum maer gael halae, moonhe na khulo halae. Khonki atae halau ke kutchu khae na pare halio hau. Saptah dus din baadekutch dost hamra kahalthin ke chuna aur haldi ke pees ke Galva mein lagvel. Teen chaar din ekdum thik bhae galae aare konho kharcha paani na kare parlae hamra.* (Cold and fever get cured within a few days. Medicines reduce the immunity and weaken the body. If you go to the doctor, they must give you some medicine. Last month, my throat was almost dead, and my voice was choked. I had a dry cough. When it got prolonged for a week, some friends suggested I use warm turmeric mixed with lime and apply the paste around my neck. Turmeric powder was available at my house. Lime was in my *chunauti* (a tiny box to keep tobacco leaves and lime, and this box is available to people habituated to '*Khaini*'). It worked as a wonder medicine. In the next 4 days, I was relieved from the cough without paying any money to the doctor)." Until he gets bedridden, Kabir cannot stay at home. The work at Nagarpalika is essential to both Hadi workers and the Nagarpalika. However, the choice to avail treatment is also governed by his socio-economic conditions. Hence, the question of Hadi not visiting health care centres has various dimensions that guide it.

In the above narrative, the reference to the throat as dead is preferably chosen to refer to the fact that Kabir was unable to eat anything. The severity of throat pain did not allow him to talk properly or even drink water. But he kept away from the clinic and medical shop. His idea was to ignore medicine as it would require some expenditure. Kabir also saved some amount by not seeking treatment and following home remedies using turmeric and lime, an old or traditional method to cope with illness. He was also aware and conscious that his illness would not spare him from reporting at work and, therefore, could delay seeking alternative methods.

The income situations influence Hadis's perceptions. However, it does not vary to a great extent among Hadis. Ramesh Hadi is one such example. He is a retired employee who has worked for more than 25 years at BGH. He worked as a sweeper-cum-helper. For a long time, he was posted in the hospital's Psychiatry ward. In the past, he had heart-related ailments and had made visits to the doctor in his hospital. It took a couple of months to be diagnosed that his heart was not functioning correctly as there was a major blockage in his heart. The disease was diagnosed at a serious stage of illness, and he was advised to get a pacemaker put in his heart. He felt his legs were shivering and were not in control after he was informed about the disease. His senses got numbed for a moment. But after several rounds of visits and consultations with the specialised doctors at the BGH, a pacemaker was implanted in his heart.

Ramesh Hadi narrates, "Mine was the first case in my community for the treatment of heart disease. I came to know that this disease is life-threatening, but it was new to the entire Hadi Cooli and to my family as well. I requested the doctor at the hospital to waive my charges as I was told that the machine would cost not less than one lakh. The patient is supposed to buy the pacemaker, and the surgery will be conducted at the BGH. It was with the grace of *Mansa Devi* (Goddess Mansa) that my repeated pleas to the surgeon brought me great relief as I was given a pacemaker free of cost."

The above narrative significantly brings certain aspects of health care services, benefits of working in the public sector, availability of special treatment and modern medical support system that helps even a retired person. As people in the district are referred in general due to the absence of Sadar Hospital, Steel Plant Hospital functions as a referral to many patients. However, most employees avail its services for minor ailments as well. But it is not a referral in principle. Therefore, a process of social construction takes place, and the perceptions are influenced by the kind of experiences a patient goes through. It is innate to have a different kind of experience in the case of a person who got a pacemaker implanted in his heart. But the patients who could not find some medicines at primary health centres are another instance. Ramesh feels that it is a wonder in his life. His entire household was in disbelief and was unable to understand how this miracle transformed the life of Ramesh Hadi and saved his life. He survived with the help of modern bio-medical technology. This could happen only because he was given medical aid from his employers, and the BGH bore the cost. Hence, he believes hospitals are the places to get appropriate health services for the poor like the Hadis. There is a government provision to provide certain medicines free of cost if a patient is below the poverty line (BPL). However, in most cases, Hadis could not get medicines in government hospitals, unlike Hadi employees in BGH. Reaching government hospitals for free medical facilities is an expensive proposition for the Hadis.

It has also been observed that not only do the Hadis lose of money on health care expenses, but they also lose additional income when they have to visit hospitals as they have to miss work on those particular days, as in the case of Kabir Hadi's narration. Pintu Hadi once mentioned that, "*Dactor to miltau kintu kharcha nae miltau* (You get doctors but not expenses.)" Further, he narrates, "Treatment will be followed, but where do we get such a huge amount? In our community, people do not have a huge amount. This is the only reason we do not go to a big private hospital."

It was also seen that there is a tendency wage not to waste even 1 day's wage, and this is possible only by avoiding the visit to PHC and Chas Sub Divisional Hospital. PHC is located at a distance of 2 km from Hadi Cooli whereas Sub-Divisional Hospital is around 5 km in the outskirts of Chas town. Hadis refer to Chas Sub Divisional Hospital (CSH) as Chas Hospital. The district hospital for Bokaro is located at Sector-1 of Steel Plant residential colony. Chas Hospital mostly caters to the needs of both rural and urban populations, including people from the adjoining Chandankiyari development block. It functions as a referral centre in the district. Simultaneously, there are quite a few private hospitals and clinics that cater to the needs of urban patients. Strategically, these private hospitals are closer to Hadi Cooli and in the centre of the town, unlike Chas Hospital, which is located on the outskirts of the town. The details of private practitioners are discussed separately in one of the sections in Chap. 5. Private health care is quite expensive to avail treatment, but where do they consult? There are fewer alternatives, and they are left to seek treatment in private hospitals.

In an interview, Ganesh Hadi revealed that he had lost almost 4 days of wages when he had to go to Chas Hospital for treatment. A few Hadis conduct their medical visits by going to the PHC and Chas Hospital. Some patients were informed that the doctor referred them to private diagnostic centres for investigations rather than at Chas Hospital. The city centre is located in Sector 4 residential colony of the city centre in Sector 4 residential colony of the Steel Plant. There are a large number of pathological laboratories, clinics doing most medical investigations and a few private hospitals and nursing homes. This market is a well-developed shopping centre in the Bokaro district that attracts most shoppers in Bokaro. The city centre market has one multiplex and a Mall that has increased the number of consumers and other clinical centres. The private practitioners and doctors in Chas, namely, Dr. Hospital, also refer their patients to city centre laboratories, namely, Dr. Lal Path Labs, Religare, We Care Scan and Diagnostic Centre. Incidentally, the doctors at the city centre at We Care Scan Centre are available in the morning hours between 10 AM and 1:30 PM, which are the working hours for Hadis. So, unless it is extremely necessary they would not want to lose work during that time.

Distance emerges as an important factor in Hadis's mind, and the decision to visit or not visit the government hospital is guided by the perception of distance. Time spent instead of 1 day's work loss and money lost as an absent day about consultation fees saved are the quick calculations Hadi make, and these are commonly understood by all the factors that also influence the perceptions of Hadis towards public health care centres.

However, a few people prefer visiting government hospitals. For instance, Nuniya, the mother of Jhomda Hadi, visits the PHC at the checkpost. She described that seeking treatment at a government hospital has some possibility to minimise the expenditure. That her narrative, she adds, "There is a lot of health expenditure in my house and with the small earnings, saving is difficult. That is why we prefer reducing the expenditure by not visiting private hospitals. In the government hospital, although we are told that the treatment is free, medicines are often out of stock, and we have to buy them from the market." She knows that the non-availability of medicines at the hospital is a common complaint. However, the meagre income of her son makes her visit a government hospital. Jhomda does not earn beyond his salary, and he has to spend a certain amount on commuting to the office in the City Centre, Sector 4. The cost of treatment and the income of the household are closely associated with each other. Illness is perceived situations differently according to the individual's income capacity to bear the expenses.

Illness and health are two concepts embedded in the same cultural phenomenon. The expressions of the individuals to feel unwell and take rest refer to a kind of illness people translate those feelings to their body as healthy or unwell and ill. The feelings are verbally explained with the help of symbols, gestures, and symptoms. For instance, Hadis express their inability to work when they feel uneasy and unable to maintain their bodily capacity to do everyday work. Leave from work would have a negative impact on negatively impact earnings, and even employers do ask to report if it is not a serious illness.

Nagraj is 22 years old and is a Hadi sweeper. He is the eldest son of Gita Devi. He lives in a joint household with his two unmarried brothers and one sister. He is employed at Hotel Naresh in Bokaro Steel City on a contractual basis. He that, "*Hamani ke kaam mein break nae ho, nae to sabke ghar break lag jito, nahii konho dharna-warna hau, bihaan aibeen to kakro nae paebeen ghara mein, kaise bardasht kare parbin hau ek diin kae harjana bharail, ab ekdumein man na karo hai kahio kintu man ke kii hai, kaam to gaud haath se hovo hai* (There is no break in our work or occupation else there will be break-down in everyone's house in the Hadi Cooli, and no food will be served. Neither will we go nor have we gone on any strike in the past. If you visit Hadi Cooli in the morning hours, you will not find any adults at home. How can one bear the loss of pay for one entire day? There are a few days when I do not feel good, but feeling unwell or not good does not have any significance; after all, work is done by hands and legs, not by feelings)."

3.4 Doctor-Patient Relationship

A doctor-patient relationship is a crucial factor in shaping perceptions of illness among the Hadis. It is therefore important to note that studying the Hadi community and their perception of illness and health requires looking at the health-seeking behaviour during illnesses and the interaction between a doctor and the patients. The widespread belief among Hadi households is that the government hospital is not made for them. This perception is due to unwarranted and undesirable experiences encountered in the past visits made to govt. hospitals to seek treatment. Nuniya Devi, a 65-year-old widow, said, "Chas Hospital is the worst hospital; no medicine, they do not speak softly and are always in bad humour. Doctors appear to be doing some favour to the patient. They scold us for no reason when we ask for medicines. I could not understand why I was being scolded in the Primary Health Centre (PHC) located at a check post in Chas town when I fell sick after a week of fever and body ache in the winter of 2008. I just wanted him to hear me out and also write to me about those medicines available in the hospital. But he was unhappy that I was asking him to give me the available medicine freely. I also told him that I cannot buy medicines, it will be a great help to a poor woman but nowhere could he show some courtesy." Although Nuniya is okay with visiting the govt. hospital, such experiences negatively affect her perception. "Sometimes a doctor may end up giving good medicines," she said. There is a sustained hope in her.

This narrative also adds another dimension to doctor-patient relationships. It also displays how and why, despite being denied proper treatment and medicines, a few patients return to the primary health centre to consult a doctor.

The above discussion also describes the discomfort of a woman who wants to visit a government hospital but is facing difficulties and discrimination in doing so. This also contributes in a significant way to them not wanting to recognise the illness as one unless it is severe. Illness narration at a doctor's consultation room is difficult for a poor Hadi. Women who are pretty vocal in their community also feel helpless when doctors do not empathise with them. Hadi women are found not to prefer to report illness than their male counterparts. This is because women are worried about expenditure and make efforts to save out of household income for the marriage of their children.

The level of trust between Hadi patients and the local doctor is high and mutually strong. They are like family members to him. Doctor Vimal Singh says that he is a pious being for the patient. The doctor admits that the treatment given to the patient is limited, unlike in big and private hospitals, namely Muskan and K M Memorial Hospital. Doctor Vimal Singh says, "Sir, we know our limitations. You would call us RMP but we function as a lifeline in small Mohallas. There is not a single MBBS doctor in the neighbourhood of as many as eight Mohallas. I do attend to patients at their homes in case of emergency. But you better know that no MBBS would be ready to serve the way we provide health care services, and they cannot match the level of our health care services by visiting even at home. We have kept alive the belief in a doctor among Hadis and others. After all, I, too, have to run my household expenses. The Government does not provide basic health care, and in every meeting, the Civil Surgeon blames RMPs and quacks as a problem to the health care systems; I don't understand any connection between government hospitals and RMPs."

Apart from the doctors (RMPs) available in the Mohallas, other doctors lack socio-cultural training and orientation; they do not connect easily with the socio-cultural background of a Hadi patient, and moreover, they treat them as isolated from the socio-cultural set-up that they live in. The doctors cannot spend a little more time with the Hadi patient, which could help in examining the family and social background of the patients while hearing the illness story. On the other hand, poor, uneducated, and lesser-educated patients have a lower capacity to explain their illness problems to a doctor in the limited time available during consultations. This aspect also turns out as the inability to be heard by the doctor is carried and shared among household members and friends in the Hadi Cooli. There are some gaps between a doctor and a patient observed during a doctor and a patient observed during the consultation that appear quite important in shaping the perceptions of Hadis. The role of sociological research data brings this gap to a minimum in the background of medical and sociological writings.

The way Hadis perceive themselves is a crucial factor that influences their view of others' perception of them. The interaction of Hadis with various health institutions plays a pivotal role in constructing their perceptions. For instance, encounters at the reception of a clinic or a hospital, either government or private, use of words during the consultation process with an attendant outside the doctor's chamber, and finally, patients' face-to-face interaction and narration of being ill to a doctor in his/her chamber, are certain levels of interactions that in turn affect the mindset of a Hadi patient. Hadis is found to be carrying dual identities: first, an identity of a patient and second, a socio-cultural identity as a Hadi who performs menial and polluted jobs.

Lamti Devi, who is 52 years old and a widow, recalls how her husband faced difficult times. She pointed out that during his life, Paltu Hadi visited Chas Hospital.

This happened seven to 8 years ago, in 2003. He had a wound in his leg that was worsening gradually. He could not walk properly but was still attending to his duty. She took him to Chas's hospital, thinking that they could save some amount apart from paying for the travel. It was not a fruitful visit because the doctor did not pay attention to his problem. He asked Paltu to meet the compounder who would clean and put the necessary ointment on the wound. Both of them requested the doctor to give them some medicine, but the doctor said that there was no need to take any medicine. Gradually, the wound worsened. However, the bandaging was done and they returned. The feeling of not getting appropriately treated by the doctor saddened the couple. The wound became a headache for Paltu. Later, after a week, he returned to a private doctor near Chandra Talkies, a cinema hall. The doctor performed a small surgery on the wound. The wound had worsened badly and could have caused Paltu serious trouble. Ten days of medication and regular dressing were done. He had to spend almost Rs. 8000 rupees for this illness. He also missed several days of work. Several investigations were conducted and an x-ray of the infected area was advised. Lamti says life was tough when her husband fell ill, as it directly affected their earnings and food intake.

For Hadis, if the sick person is the sole earner, as in Lamti's case, the situation worsens. "The doctors at the government hospital do not show any sympathy towards the poor," she admits. She continues, "If they get proper suggestions and positive behaviour from the doctors at the hospital, poor households like hers can survive and add something to household earnings, but this is not going to happen in our lives. The poor are more humiliated."

The public health understanding put forth in the studies (Qadeer 1995; Qadeer and Dunu 1989; Zurbrigg 1984; Banerji 1971; Banerji 1978; Banerji 1984; Sahu 1986) provides a perspective to look at the perception of Hadis towards the availability of health care services. In the above-mentioned studies, three of them, Banerjee, Qadeer, and Sahu, have discussed the need for public health systems in a country of unequal society such as India—large numbers of the poor demand a promising public health system and its implementation.

Zurbrigg (1984) gives a qualitative account and description of illness and accessibility to health care treatment faced by Rakku. The poor financial conditions of Rakku tell the ill health experiences of the poorest households, and the death of a child is not one reason but is connected to a chain of socio-economic and political factors. It is important to see how the questions raised in the story of Rakku lead to the analytical steps. Rakku's thatched mud home is located far from the village, without any doctor in the hospital. She took her child in the last stage when a woman did not find health care support. It shows the very structure of Indian society that does not help Rakku to save her child.

From this perspective, "Poverty is seen both as the primary source of ill-health and as a force which renders the poor majority powerless to make effective use of the official health system, and powerless also to change it" (Ziyauddin 2022, pp. 14–15). Some of the above-mentioned studies have taken the line of argument of the Bhore Committee (1946). They argued that good financial support should be allocated, and the focus should be on strengthening primary health care services and not only putting high budgets in urban hospitals.

In her paper on 'Draft National Health Policy 2001-I: Debt Payment and Devaluing Elements of Public Health', Qadeer (2002) highlights the undermining of the referral system in India's new national health policy, 2001. Therefore, the point made in the above studies is that public health provisions not only help in reducing the expenditures on health for the poor but also that there is a high expectation among the poorest for public health care. They seek low-cost treatment. However, it all depends on the kind of services available at government hospitals. Women like Nuniya Hadi overcome their discomfort and prefer to ignore the indifferent interactions with doctors and seek treatment at a PHC. The inability to pay for private consultations and inaccessibility in government hospitals put Hadis in a difficult situation.

Moreover, the lack of capacity to pay in private hospitals also explains why Hadis turn to RMPs as their preferred point of seeking treatment. In a paper titled 'Social and cultural foundations of health services systems', the analysis of Banerji stands correct in relationship to the growth of RMPs and the failure of primary health centres to reach poor either in rural or urban. He writes that "The availability of medical health services and capacity of patients to meet the expenses are the two major constraining factors" (Banerji 1974, p. 1341). One of the arguments put forth in his paper is that the outnumbered population has not been given an equal level of medical care service. It has left them to depend upon RMPs. This shows the unmet need for medical care services for the poor.

Most RMPs are present in the absence of government health institutions. A good number of RMPs have taken place at the first healing centre. For thousands of people in localities neighbouring Hadi Cooli, public health institutions function too far away, whereas there are three RMPs within a half-kilometre radius.

The increased population has also increased the demand for health care in the urban Chas, and the deliverance of the service has not matched the needs of the rising town population. As per the Census population records, the population of Chas town was 97,221 in 2011 which increased from 1 to 41,618, a total of 1.6% of the total state's urban population (Census of India 2001; Census of India 2011). Private health care providers have established their foothold among the urban population, whereas Registered Medical Practitioners (RMPs) have set up clinics in poor localities in the town of Chas, two of which are in the neighbouring vicinity of Hadi Cooli.

Doctor Vimal Singh is the most closely located at 5 min of walking distance, making his clinic the most convenient treatment centre for Hadis. For poor patients like Hadis, the cost of travel deters them from travelling to government hospitals and forces them towards private clinics, even when a few thinks to visit. Perceptions of Hadis towards health institutions are also influenced, like in the above context, by distance, cost, and convenience.

The role of RMPs becomes more important in small urban settlements with poor households. They are familiar with one another. Doctor Vimal Singh has been practising for over two decades told that, "I am not being differentiated as untrained." Rather he is called 'doctor' by all his patients including Hadis. The only inference is *baed dactor* (bigdoctor) and *choet dactor* (small doctor). Hadis refer to

RMPs as a doctor and those in big hospital are called the big doctors equipped with better equipments and investigations. When Doctor Vimal Singh refers a patient with serious illness beyond his diagnosis, Hadis feel that he referred to a bigger doctor or a hospital. Differences are not absolutely clearly available, but most Hadis have seen private hospitals like Muskan Hospital, K M Memorial Hospital, and Neelam Nursing Hospital, and they find certain differences. The clinical set-up, charges, and doctor's assistant are a few indicators that emerge as differences between doctors in their locality in comparison to other health providers, as mentioned earlier.

While Hadis preferred to visit the RMP doctor in Bauri Cooli, the alternatives are limited. Dependence on local doctors is not chosen but driven by the fact that public health care fails to provide for poor populations like Hadis, and they cannot afford to pay the fees of qualified medical practitioners.

In a first impression, Hadis were found to be choosing RMPs as their first referral, showing higher utilisation of health care at Doctor Vimal Singh's clinic. Patients have also been going to government hospitals and nursing homes for child delivery-related services. Simultaneously, Hadis have also lost many traditional healing practices and have limited healing traditions surviving in the community. The practice of RMPs as doctors is most popular for seeking and providing health care until the illness is quite serious and untreatable by the RMPs.

The practice of allopathic medicine is rampant, but the costs vary. There is no standard doctor's fee as the fees vary per patient, their earnings, and the rapport of the households with the practitioner. Doctor Vimal Singh said, "I charge 20 rupees from some patients, and there are patients from whom I charge Rs. 150 rupees." The fee also varies based on the treatment. Suppose the patient is given an injection and a few medicines in an emergency. In that case, in an emergency, the fee increases up to rupees Rs. 200 and maybe more. The fee paid varies in the case of a home visit. During a home visit, the charges almost double at 200 to 300 rupees. There are several cases (patients are referred to as cases by RMPs) who cannot walk to the clinic. The doctors are then requested to visit the patient and provide appropriate treatment. Doctor Vimal Singh said, "In case of a home visit, the fee depends on the disease a patient suffers from. The fees also include time taken for visiting. I have to close the clinic or ask someone to inform the other patients that I am away for a home visit. Home visit is also the reason I, or any other doctor in the locality, may lose other patients. The poor Hadis are mostly suffering from some health problems or the other. They need immediate care after delayed reporting. It is a support to them. Their immunity is low and they catch infections from others easily. Immediate relief is of prime concern to me and the household members as well. Hence, I have a better scope to earn a decent amount, and people pay happily and with full trust in my treatment."

Doctor Vimal Singh finds the level of health status very low in Hadis. The frequency of visiting his clinic is higher among Hadis and Bauris than among other castes. He feels it is their incapability to eat nutrient-rich food, such as milk and vegetables, unlike other caste groups. This is his observation after years of interaction with his patients from Hadi Cooli, who are into filthy occupations.

Caste-based social distancing also affects the doctor-patient relationship, and a few Hadi members, but not all, experienced this. Other castes also socially distanced Hadis. For example, Pradeep Kumar is the son of Somesh Hadi.

He is a 15-year-old student in class 10. He says that they are neither polluted nor outcasts. He has friends outside his caste and visits their homes. It is through schooling that he can make some friends outside his caste. One of his school friends, Rahman, is from Telgadia village, located within the Chas Block area. He plays with him whenever he visits *Telgadia* village. Many of his peers do not agree with their elders on the issue of discrimination and social distance. Hadi youth are well-dressed and wear clean clothes after returning from work. Most working youth change their clothes in the afternoon hours after sweeping and cleaning. Pradeep says that his father has told them the stories of discrimination and humiliation in his early days of childhood. It was terrible to hear the details.

Pradeep narrates, "Yes, the discrimination has reduced much now. But if we open up and tell our identity to outsiders from other castes, their facial expression changes momentarily. It happens whether in a hospital or anywhere else. In my school, I have been treated unequally on a few occasions. In the football team selection at my school, Ram Rudra High School, I was kept out of the team despite my performance as a defender. It is a different matter that I play regularly on the football team. My father can tell you so many stories." It appears that his school team did not take him in the inter-school tournament, which made him feel discriminated against. But other students also felt they were not given a chance. The possibility of a caste association may be one of the reasons as football team selection is based on performance in terms of winning ability factor.

Pradeep further says, "How to understand the others? Sir, people are nice from the outside, but you never know what they think in their mind about others like me, especially the Hadis. Our parents are very simple and have not changed over some time in terms of dress and outlook. We do wear new clothes and clean clothes, but they are still dressed like older times or sometimes in dirty clothes. They are habituated to that but want their children to do well. However, I know that my caste will continue to remain Hadi. Changing our caste is not possible, but we can bring some change by going to school and doing well in our studies. But most of us are not interested in studies which could help to change our work status if not caste. I am in my last phase of schooling in a government school at Jodhadih More, Chas."

His father, Somesh Hadi, narrates, "Generally, people do not prefer to sit with us if they get to know that our profession is that of sweepers and we are Hadis. You know, I do not visit the government hospital as I have a health card for Bokaro General Hospital as a retired employee. However, the issue of discrimination of Hadis emerges somewhere or the other. If I visit Doctor Vimal Singh or Ganesh, they will not inquire about my background or profession, nor will they be bothered about my area of residence. It is known to them already. They would enquire about the illness and raise questions related to the illness. They understand our health problems because they understand us and are familiar with our lifestyle, so unnecessary questions are not asked. The local doctor does not create a barrier between the patients and doctor."

Somesh finds that some social distancing is practised in hospitals outside his neighbourhood. It is with Doctor Vimal Singh or Doctor Ganesh that he does not find himself discriminated against. Most Hadis perceive that the doctor-patient relationship is always unequal, the doctor finds it placed higher in hierarchy and knowledge than the doctor finds it placed higher in hierarchy and knowledge than the patient. This factor also governs all consultations sought by a Hadi patient. Doctor Vimal Singh has been familiar with Hadis for quite a long time. The relationship is beyond a formal interaction that takes place in a clinic. Greetings and encounters are also possible out of the clinic when they meet each other outside of the hospital, which is barely possible with any doctors practising in hospitals and clinics. The doctors in the neighbourhood reside in another nearby locality called Shakaro Cooli, which makes informal talks possible other than the clinic.

Recalling her experiences, Nuniya Devi discussed another story. She said it is difficult to talk to the doctors during a visit to the hospital. She found it difficult to interact and share her illness at Chas PHC in the early months of 2010. She says, "He (the doctor) does not even ask me properly. When I asked him to check what kind of illness I was suffering from and why I felt so weak and fragile, why I could not sleep and also lost my appetite. I was still explaining my problems, but the doctor said there was nothing to worry about and it is common to all aged persons. He wrote two to three medicines and asked me to buy them. Afterwards, I just couldn't ask him what exactly was wrong with me and if I would be cured! "No, there is nothing; you go now," he said, directing me. When I stood and started going out, I thought of asking him again but, the moment I turned back, he said, in a loud voice, "Go now!"

Ill experiences leave some impact on the mind of a Hadi, perhaps because their caste, occupation, and poverty sometimes bring scolds and abuses. Nuniya could not understand the interaction with the doctor in his chamber. She felt a little upset. The louder voice and impatient hearing of the illness narration are a few aspects she found to be annoying her and affecting her mindset in the course of seeking health care.

Better diagnosis and treatment are possible when the hospital gives space and respect to Hadi individuals; until then, Hadis would not believe in physicians or hospital administration. This analysis can be supported by the narratives of Nuniya as a patient. The non-availability of doctors at the hospital on time and the non-availability of medicines and medical investigations minimise the visits of Hadis to public health centres, and the relationship between doctors and patients is strained.

On the question of whether the discrimination is perceived equally among all the Hadis in the vicinity, a group discussion was held at Hadi Cooli during the fieldwork in January 2011 with Hadis who joined were Somesh, Ramesh, Pintu, Komda, and Ganesh. I tried to find out why most Hadis do not like to avail themselves of health care at the PHC and Chas Sub Divisional Hospital but rather prefer to avail themselves of private health care services. A few cases were discussed that showed the field situation to be complex. For example, Komda Hadi is 30 years old and never believed in the treatment provided at the government hospital, Chas. He believes that visiting government hospitals wastes time, and, consequently, 1 day's wage is lost too. The Outpatient Department (OPD) is in the morning when it is time for Hadis to go to work in the first half. When they come back in the late afternoon, it is late. The OPD timings and working hours of Hadi sweepers clash. Komda knew that when his mother went to the hospital at the checkpoint, she did not get proper treatment. She told him that it was too far to walk down and that there was no support for the poor or any medicine available. Komda Hadi never went to Chas Hospital due to his strong perception that private doctors are better than those in the government hospital.

There is some convenience of visiting a private doctor that suits him according to his work schedule. He said, "You go and ask anybody in my community; no one will have good words about government hospitals due to their experiences. If the government had helped us, we would not have been living like this. We do not benefit from government schemes and planning (*Yojna*). I cannot go and take some government benefit at the Block office. I am not aware whether the government provides support to Hadis or not. Unlike us, at least Bauris have a presence and bargaining power during ward elections. Their numbers are larger than ours. I am rarely interested in visiting the government hospital. Our doctor in the locality charges 50–100 rupees, including medicines. Even when we visit in the evening hours, he is available, or in case of a sudden need in the night, he welcomes us and provides treatment. Both my children have had regular bouts of asthma. Deepa, who is 11 years old, has had so much trouble that I do not want to recall her troubles. I was running from pillar to post for almost 6 months. I even visited Purulia town, a neighbouring district in West Bengal, to meet a traditional healer, Prasanjit. We call him Ojha.[3] The Ojha did some *Jadu-tona* (witchcraft) and performed rituals to cure my daughter. My daughter had breathing difficulties at night when in bed. She would not sleep the entire night due to breathlessness and discomfort. In such difficulties, how could both of us sleep? Now I cannot tell you whether medicines or Ojha treated Sonam. At times, I feel that the Ojha brought this wonder into her life. But at the same time, we think that maybe both medicines and Ojha helped to cure her illness. She is fine now."

Before visiting the Ojha, Sonam was taken to Doctor Vimal Singh and later to Doctor Dasgupta. Doctor Dasgupta has a clinic located at Cooperative Colony, around 8 km away. Under his treatment, her illness improved slightly, but it continued troubling her at night. It was almost 3 months' time when Komda decided to meet the Ojha.

For example, Hadis find themselves in confusing circumstances when meeting doctors at hospitals or health centres. Ganesh Hadi narrated his experience at PHC of check post locality in Chas town and said, "I had extreme pain in my stomach, I was unable to eat proper food and was not at all feeling comfortable. The Doctor did not give a patient hearing my complaint of stomach pain and feeling of an impending illness. The doctor immediately wrote five to six medicines and asked me to leave. While I was leaving, the doctor asked me why am I acting so much; there was nothing to worry about; go now, will be all right." There is another story that describes the experiences of the illness.

---

[3] Ojha is a person having knowledge of illness and cures it with his or her witchcraft and certain rituals, in the region of Bokaro and surrounding districts.

Ganesh Hadi's daughter, Anchal Kumari, aged 11, had severe pain in her stomach. She has been suffering from stomach pain for the last 2 years. Her pain would automatically subside in a day or two. Some medicine from Doctor Vimal Singh in the locality would help the pain subside temporarily, but it never got permanently cured. Vimal Singh advised Ganesh to visit the sub-divisional hospital at Chas to take advantage of the concessions given for various tests and investigations. He narrates, "I took my daughter to Chas hospital. I had never been there before. As I learnt, they hardly provide or prescribe effective medicines and we do not get proper treatment as well. I went last winter (2009) at the suggestion of our doctor, Vimal Singh. I reserved an auto rickshaw for Rs. 50 rupees and took my daughter early in the morning to the hospital near ITI more of Chas town. It is on the outskirts towards Purulia Road. After 3 h, we met the doctor as he arrived late and spent less than 5 min with us. He did not enquire and listen to Anchal carefully. He just enquired what the illness was and for how long it had been there. We skipped our breakfast so that we could reach it on time. In 2–3 min, he asked me to take my daughter for the prescribed medical investigations. He did not give any medicine. I asked the doctor where and how we should get these tests done. You did not give any medicine; how do you expect the pain to subside? But only a one-line response was given. He directed me to ask the compounder, who was supposed to explain the details. Nevertheless, when we looked outside the chamber, the compounder was not around.

When we met the compounder after lunch. Hence, the tests cannot; he said the machines were not functional. Hence the tests could not be done. All these tests can be done only at the city centre in Hi-Tech Diagnostic Centre, near Shivam Hospital in Bokaro Steel City." Ganesh Said, "The next day, I went along with Anchal to the city centre of Bokaro Steel City colony. We took a shared vehicle to minimise our travelling costs. The City centre is located at a distance of around 10 km from Hadi Cooli. After reaching the hospital, I put my Card to get a number in the queue. The blood and urine samples were taken but the X-ray got delayed. The technician had gone for prayers as they go every Friday. Once he returned, the X-ray was taken. But during this process, he asked me why I wanted to go back to Chas Hospital when a doctor was available in this very hospital. He also told me that this doctor is a famous one. He charges 400 rupees and if he suggests further investigation, the technician promised to help me get some concession. I was convinced after listening to him. The doctor did not write further investigations, but the expenses on the entire treatment crossed four to five thousand, excluding the travelling expenditures. After a few days, blood samples, X-ray and ultrasound were again conducted along with MRI."

The situation takes a handsome amount incurred on medical investigations. He said, "All my savings got drained in this process. There was nothing left in my pocket that month. I cannot recall the travelling expenses now. What more can I tell you, dear? I do not want to recall and think about past tragedies. Since you started asking, this incident came to my mind. What is the need to remember such problems when there is a new daily challenge in our lives? In a couple of months, we get unwanted troubles and, in the process, lose the entire savings or whatever is earned. The cost of treatment by private doctors or hospitals is always higher. But the matter of the fact is that my daughter Anchal is healthy now. There was some trouble in her chest as it was found at the city centre hospital. But now she is fine and is able to carry out all her routine work. She also helps with household chores."

The above narration talks about access to health care services and treatment sought for an illness that prolonged the patient's suffering. The pain and discomfort not only troubled her every day but also cost the family a lot. The data reflects another aspect of the treatment-seeking process when the expenditure crosses the limit of the household's capacity. It also addresses a public health institution's functioning that does not provide certain investigations resulting in an extra burden on poor patients. The non-availability of medical investigations is a kind of burden to the poor patient. But in the case of Anchal's prolonged illness, it was suggested that she undergo certain investigations. Her father, Ganesh, found solace in that the doctor treated her properly.

The above narrative also brings a new dimension to the research. If the doctor in the locality refers a Hadi patient, they abide by him and attempt to seek proper treatment outside Hadi Cooli. A doctor-patient relationship develops a kind of trust. They understand each other's situation. In these situations, the doctor also talks about various aspects of his patients, not only illness.

As discussed above, Hadis did not hesitate to do the additional job of removing dead animals if found in the city area. It was a way of earning extra bucks. The point highlighted in earlier sentences is about the change in the available jobs. The memory that dead animals were to be eaten is also brought up in interviews with Elder Hadis. Somesh says it was not prohibited to eat dead goats, poultry, etc. if the meat was not damaged. He said, "It was not considered demeaning and filthy to eat such meat. It is as if a family would get meat free of cost and have a feast for a day. You know, now there is no practice to prefer such meat at present."

Ramesh says, "When other castes do not eat" '*moer janavar ke maans*' carrion meat; even Muslims never eat such meats, then why should we eat? The meat smells bad." It is justified that now, no one likes to eat the meat of dead animals. If our doctor hears someone eating such meat, he will scold them. I also served at the houses of a few officers in the steel plant colony of Sector-4 in B. S. City. At times, they asked me whether Hadi still eats meat from dead animals, and I would say that we have completely left eating it. It was a practice of the past in earlier generations. Now Hadis have changed and find it disgusting. I wonder how we ate such meat; I get frightened and feel nauseated now. Removing and throwing away the carcass is largely all right; four people would pick it up on their shoulders and throw it away in the outskirts of Cheera Chas (there was largely open forest land). Now, it is an almost 20-year-old story; life has changed a lot in this period. Though Hadis has not prospered, unlike other castes of the area, the food patterns and habits have changed among our households. Now we also buy meat from the market whenever we can bear the expenses. Vegetables and roti are commonly preferred among most of the young Hadis in their households.

He added, "I am not familiar with the herbs they used. Young children study and go outside and interact with people other than in Hadi Cooli; they have better knowledge about life outside Cooli. They do not want to listen to the elders, and the impact of television is such that they do not have the patience to listen and feel more self-confident than the elders." Various changes are increasingly felt in several urban wards, including the Hadi Cooli ward. These changes are recorded in the following two points: firstly, television has impacted the lives of Hadis, causing a shift in food habits and ways of dressing from unhygienic to more clean and acceptable ways of living and secondly, local doctors also seem to impact various food and cleanliness habits of the community. The food pattern has seen some changes. Many young Hadis said that they like wheat rotis rather than boiled rice (*Maad-Bhaat*). Rice is an economical food, but roti is preferred. However, the older members give first preference to rice and curry (potato, tomato, vegetables, etc.) The doctors keep suggesting that Hadis eat hygienic food and dress in clean clothes. The information coming out from a doctor again increases the trust between Hadis and the doctor.

Emerging Views on Food Habits:

1. Television has impacted the lives of Hadis, causing a shift in food habits and ways of dressing from unhygienic to cleaner and more acceptable.
2. Local doctors also seem to impact the community's various food and cleanliness habits.

3.5 Narratives of Prejudice in Seeking Health Facilities

Early (1982) states that patients' narrative gives voice to the suffering in the form of a therapeutic narrative. This was further classified by Hyden (1997) into three types of illness narratives: Illness as narrative, narrative about illness, and narrative as illness. In Illness as narrative, narrator (mainly patient), illness and narrative are combined in one, and the illness is expressed and articulated in and through a narrative. Kleinman says that narrative plays a central role in the illness's occurrence and in shaping the manner in which it impinges on the individual's life. The illness narrative is a story of the patient. Significant others retell, and significant others retell to give coherence to the distinctive events and long-term course of suffering. The narrative does not merely reflect the illness experience. Rather, it contributes to the experiences of symptoms and suffering (Kleinman 1988, p. 49).

For Hadis, validation of expertise, usually concretised through explanations sought for both illness and cure, is a particularly important crucial factor that determines their visits to the health care provider. In other words, the time given in the process of interaction with the health care provider to prescribe a cure is essential to patients. Narrating an illness demands at least one listener who listens to the feelings of being unwell. Hadis see the indifference of health care providers as a caste-based prejudice.

Ramesh adds visit to that the association with the occupation of manual scavenging creates several barriers during visits to health centres and hospitals. The social distance between Hadis and hospital staff is not only felt by public health providers but also by private health care services to an extent. However, the severity and intensity of discrimination vary between the two. Private caregivers present fewer manifestations of social discrimination towards Hadis than public health services. It is also found that private doctors care least about caste and occupation; they look for patients who can pay fees and investigation costs. To a private practitioner, a patient's visit is important. Neelam Nursing Hospital is mostly visited for issues related to pregnancy and delivery matters. The nursing staff do not behave nicely to Hadis. The cost of treatment in private hospitals is another deterrent for poor Hadis. However, most Hadis consult at Neelam Nursing Hospital if the pregnancy and birth are complicated.

Several respondents made these observations during their lifetime while seeking health care. Jhomdai Devi, a 47-year-old widow and sister of Ganesh Hadi, says, "When my husband was in his last stages of life due to severe chest complications, we wanted to take him to a big hospital, but financial constraints stopped us initially. How could we take Sipeta Hadi to a big hospital when we did not even have a thousand rupees to take him to BGH? We went to Chas Hospital on several occasions in the hope of getting treatment for my husband and seeking proper health care at a low cost, but it did not work out; we did not know anyone who could help in consulting at BGH. The facilities of treatment at BGH are for employees and their dependents. At times, I used to feel like punching the doctor's face. The government doctors do not bother to listen to us. My husband had a problem in his chest problem with his chest, which we could not understand. The doctors never explained the seriousness."

In the course of seeking treatment, they went to many doctors, but they never got a proper diagnosis and treatment. However, one doctor at Chas Hospital suggested taking him to BGH. In 2005, BGH would admit private patients if they had an employee as their guarantor. A person came forward at Bye Pass Road clinic and offered his support in return for 20,000 rupees. He said that the amount includes both treatment and his commission. However, we consulted Doctors for my husband. He did some tests like X-rays and blood examinations and prescribed eight kinds of medicines. But Sipeta did not survive and passed away within 6 months. Nobody wishes to go to Chas Hospital, and if you have had one bad experience, you will never revisit the place. Everyone in the hospital is running here and there. Then who will bother to attend to Hadis' illness?."

Health care personnel in general, in general, make references that Hadis are ignorant and do not want to establish social interaction outside their own community. During one visit to the civil surgeon's office in Sadar Hospital, a doctor said, "Hadis do not want to live healthy and hygienically; how hospitals and doctors are responsible for their illness." But in a short while, Doctor Ram contradicts his colleague and says, "There is a lack of health care treatment on our part. When we do not hear the illness details or take an interest in curing the illness, patients' feelings are not good. They return with hopelessness; I have seen when I was posted in PHC."

The dominant perception was that it was the community behaviour of Hadis rather than the prejudice and stereotyping by other castes and communities. It was observed during fieldwork that I am the only outsider visiting the locality of Hadis.[4] Quite a few

[4] During my entire fieldwork duration, I was one of the few who went inside their street and houses. I sat comfortably on the cemented seat (*peeddha*) made outside the houses. It was quite common

also told me that other caste members dislike visiting Hadi Cooli. This trend clearly shows the social distance between the Hadi community and other social groups in the Chas area.

It is found in the above data that a doctor also associates doctors Vimal Singh and Ganesh to the caste background and the patients visiting the clinic. Doctor Singh does not have patients from Hadis, and quite a few are from other SCs. It also reflects the caste dimension. The caste dimension is perhaps important to Hadis when they visit a doctor, as highlighted in the above data. However, it has not been found quite clearly in the data. Rather, language, familiarity, and comfort in approaching the doctor are found to be better at Vimal Singh's clinic than at Doctor Singh's clinic. In the final observation, the doctor finds it comfortable with his patients who have been visiting him. He already has a job that makes him less worried about attracting large numbers of patients. He is also not a resident of the neighbourhood locality but lives in another colony of Chas near Check Post, located around 3 km away.

Ashish Hadi, who is 65 years old now, narrates how difficult it is when someone is unable to work. He says, "When I am unable to work and am not earning, a feeling of helplessness emerges within me. Who would give me even a rupee as help? I have to feed my household members, several relatives, and other visitors at home. All these feelings arouse my sorrow and suffering during my illness when there is no work or when I am unable to do any work. Hospitals merely provide medicines. Please do not ask about the hospital staff; they are difficult to deal with." Work is essential for the household. Visiting any doctor requires some amount to be kept in the pocket to pay fees and buy necessary medicines. Ashish is a retired employee. He has a son who works at the State Bank of India (SBI) in the Chas branch. His earnings would also help save some money for his granddaughter's marriage.

Certainly, it is difficult to stay away from work and lose 1 day's wages as a consequence. The poor living condition of Hadis shape their mindset and influence their perception towards illness and coping mechanisms. Several households report that there is a trend to avoid seeking health care at the initial stage of illness. People perceive that minor breakdowns are common to everyone, and one should not get worried. Ashish also says, "Everyone faces minor illness, and at times, it gets better without medicines. I am prone to cough and fever, but I do not visit a doctor for every such illness. Our body should develop immunity and capacity like animals (*daangar*/ cattle)."

In the process of seeking a cure, there is an attempt to delay the visit. Narrating the illness is an important way to find a proper cure. Hadis do not find much time given to them to bring out the feelings in illnesses. The experiences of Hadis in a situation where doctors do not provide comfort and time are highly dissatisfying. This kind of satisfaction can be achieved by properly narrating illness to a doctor, whether in clinics or hospitals. The effectiveness of RMPs like Doctor Vimal Singh,

---

to sit and talk with them anywhere in the Hadi Cooli. But I did not see outsiders coming at any point. May be I have my compulsion to carry out the fieldwork and others do not have any need to be there.

which is judged as positive, is important as he understands the need to listen to the illness experiences of Hadis.

The proximity and interaction with the patient provide him with in-depth information. The doctor has to maintain a good rapport with the patients and maintain some good relationships with the patients, so they keep visiting him in the future as well. It is important for the RMPs to follow the requests made by patients and stay in the good books of Hadi households. Secondly, it is pertinent to differentiate the perception constructed during visits to the clinics of doctors' clinics and simultaneously visiting government hospitals or primary health centres.

The relationship between Hadi and the physician at the primary health centre is not a comfortable one, unlike the setting at a local doctor's clinic in the locality. There is less time given at a hospital, which minimises the scope of Hadis to narrate the details of the illness. If we consider the visit of Samina Devi to Dr. Vimal Singh's clinic discussed earlier, she does not hesitate to communicate with him. She did not take anyone to accompany her. The comfort while narrating her illness could be observed on her face. The space provided by Dr. Vimal Singh is never expected at any clinic outside the locality, either at a private or a government hospital. It is noted that the RMP develops a fairly good understanding of his patients, their behaviour and psychology over a period of time through many interactions and practice.

3.6 Encounters at the Medical Shop (Davaae Dukaan)

Hadis have various ways of coping with minor illnesses. First, the health-seeking initiative is also decided according to the status and role of the person in the household. The person without work is discouraged from availing of health care until illness begins to trouble food intake and physical ability to do daily chores. Second, health care is considered important for the household head or the person who is earning to sustain the household. His or her illness is directly dependent on everyday earnings and wages. It is also significant that the adult members delay in reporting illness even though there is some preference.

In the interviews with Jorni Devi and her husband Abhishek Hadi, it was found that they do not compromise on their health problems. The condition of their life is such that it does not allow them to rest. They are determined to fight minor illnesses like weakness, fever, and cold while continuing to work. Jorni narrates, "Both of us, husband and wife, have grown old. I hope that none of us should go to bed due to illness. Who would look after an old man and an old woman? Our three sons are married, have children, and are engrossed in their household affairs. Both of us work to earn our livelihood. You know, it is work that helps keep us physically active. While working, 1 day, our life will come to an end. We feel weak sometimes, but I know it is due to our ageing process and not due to any major illness. Doctor Ganesh often comes to our home and enquires whether our health is alright. In case we need some medicine, he provides."

Abhishek and Jorni Devi have convinced themselves not to fall ill as that will increase the burden on household expenditure. However, they also seek treatment whenever needed at the clinic of Doctor Vimal Singh or Doctor Ganesh. There are occasions when Jorni Devi goes to the *davaee dukaan* (medical shop) of Suhail and describes her illness symptoms to him. Visiting the medical shop helps in minimising the expenses and she finds temporary relief. A doctor makes routine rounds in Hadi Cooli at least once in a week and sometimes when he visits a patient's home for treatment. He at least tries to meet most of the old men and women and probes into their health.

Visit to a home also depends on the rush at his clinic. Vimal Singh said, "Home visits are made when no patients are in the clinic. We cannot leave a patient to wait in the clinic. They may not come to my clinic if I do not respect them. There is no dearth of private doctors (RMPs) in Mohallas. The encounters at the medical shop are less hierarchical between the druggists on the one side and the Hadis on the other. The medicines are sold on demand. For instance, one tablet for body ache is given for two rupees. Combiflam helps in subsiding body ache and fever. Diclofenac composition is also prescribed for body pain as an analgesic. Diclofenac decreases certain substances in the body that cause inflammation and pain. Symptoms related to body ache, pains, etc., are commonly reported to the medical store and, at the most, to Doctor Vimal Singh. Acigesic, Adiflam, or Acefon Plus are some of the tablets prescribed to a Hadi during my visits to the clinic, whereas Ibuprofen is also regularly given at the medical store. The most reported illnesses at the medical store are body ache, headache, weakness, loose motions, cold, and fever. A girl from Hadi Cooli visits the shop and tells the medical shop owner that, "*Hamar maae ke baes na lago hai. Kahalkau kamzori hai aare gauda mein tani peeda de hai.* (My mother asked me to buy some medicine. She is not feeling good and has pain in her legs)." After this, she puts a couple of coins on the counter. The medical shop owner asks her, "You got just eight rupees in hand, my dear; at least 15 rupees is required, and don't you have any more?" Upon which the girl replies, "No, I really do not have any more coins; she (my mother) gave me just eight. Please provide some medicines for this amount."

The medical shop owner thinks for a moment and says okay, take some medicine; she will be all right. He gives four Nabuflam tablets and two tablets of Vitamin D to help the body. He also suggests her medication schedule to take half a tablet twice for 4 days and the other one with Vitamin-D composition once in 2 days. The girl nods to say she has understood and will explain the same to her mother. Medical shop owner also added that, "*Aare taka leke aibin to ek davaee (marham) dabo okar lagve partau; Bees taka okar daam hau.* (if you come with 22 rupees then I will give you an ointment to use in the painful area of leg or hand)."

Such conversations between Hadi and *davaee dukaan* on illness provide another dimension of seeking a cure. Medicines are sought quite easily, and the medical shop owner finds no difficulty in understanding the little girl. He interacts affectionately with her, and the girl does not find any discomfort. The girl does not find any discomfort and feels uncomfortable talking to him. Suhail is a key informant in the field. He has been running a medical shop known as 'Chas Medical' for the last 25 years just after finishing his intermediate. He lives in Ansari Mohalla, a neighbouring locality of Hadi Cooli; Ansari Mohalla and his family have been residing for generations and are natives of Chas, not migrants. As his medical shop is very old, most families in the surrounding Mohallas are familiar with him. Suhail, the shop owner, also knows the Khortha[5] and Bengali languages and they are frequently used interchangeably.

On most occasions, Hadis comes with a certain amount and hands it over to Suhail, the medical shop representative. Simultaneously, they explain their feeling, problems, and illness. Suhail is also quick to decide which medicine matches both the money given and the illness. The medicine is dependent on the money in hand. Suhail says, "Both are happy in this way. When Hadis or others visit my shop, they ask for medicines according to their choice and capacity. This is the most difficult task for me. I have been in this business for more than 20 years now. I know very well about the medicines and their usage. I also understand the patients visiting my shop. Also, when there are new medicines, the medical representatives explain the medicines and their usage to us. More, this profession is all about learning by practice rather than education. I have been managing this shop since I was pursuing my graduation. Initially, there were just two doctors in the entire Mohallas (among eight Mohallas) and only one medical store, which is my shop. But now, almost four medical shops have opened in the locality. Most of the residents living in the area know about my medical shop. There are several households whose postal address is my shop. The postman still asks about those houses and individuals he cannot locate in the Muslim Mohalla. All this shows the popularity of my shop. I still have a good number of patients turning up to buy medicines. Now, Doctor Vimal Singh has helped a person to open up a new shop, and he refers his patients to that shop. Leave this matter now. We will discuss it later."

For Hadis, it is the most convenient way to explain their illness at a medical shop. Similarly, the medical shop owner also reacts positively to the Hadis. It appears that the person at the medical shop acts as a healer for Hadi people patients. There are a couple of points that make Hadis frequent medical stores to seek treatment. First, the druggist or medical shop owners, especially of the two shops near Hadi Cooli, are familiar with the Hadis. One belongs to Suhail, and the other is opened opposite Vimal Singh's clinic. Second, Suhail knows most of the Hadis by name and face. Third, the language (Khortha) of Hadis is similar and familiar to Suhail, which does not create any barrier between the two. Lastly, *davaee* (tablets) are given as per the needs of the patients and the paying capacity at the counter; lastly, the illness is narrated and heard. There is a social bond between locals and *davaee dukaan*. The social bond is culturally shared and socially important. The relationship is not only at the medical shop but even outside.

[5] Khortha is a blend of Magahi and Bengali and spoken in the districts of Bokaro, Dhanbad, Hazaribagh, Ramgarh, Koderma, Giridih, and Chatra of Jharkhand state

3.7 Experiences at Clinical Set-Ups

The perception of Hadis towards illness and health is constructed not in isolation but as an outcome of their interaction with health institutions at large, as seen in the above narrative of Jhomda Devi regarding Chas Hospital. Opinions arising from personal experiences are then shared among the community, contributing to the formation of the perceptions.

In an encounter when in service at BGH, Somesh Hadi narrates the incident which shocked him. He says, "The incident is of when my son Mangru was probably 15 years old and had fallen ill. He was suffering from a high fever. We covered him with four to five quilts (*gendra*) and whatever was available at home. Many people were similarly feeling severe cold in Hadi Cooli at that time. We took him to BGH's emergency ward at midnight. My wife Roshan accompanied us. BGH provides free health care to its permanent employees and dependents."

The hospital staff did not know he was a BSL employee. He did not carry his Identity Card in his anxious state of mind. But after an hour of pleading, he could find help come forth and was finally gratified. The hospital admitted his son Mangru for 5 days. Blood investigations were conducted. He would get to the hospital after attending duty, which was near to the hospital. The doctors would hear his queries, but none would respond properly and explain what had happened to Mangru. Somesh admits, "It is a fact that BGH saved the life of my son and had I not been employed there, God knows what would have happened." In this narrative, the care provided by BGH best supports its employees. Show an employee identity card, but Somesh Hadi was not aware of that and had no such experience in the past.

Despite being an employee, the hospital did not care to help out Somesh Hadi due to noncompliance with the admission rule in producing the employee card. But Somesh Hadi took the behaviour as unwarranted. Some courtesy should be shown to an employee. No psychological support from the doctors and staff was posted in the emergency. Such incidences tell us about the behaviour of the health care providers or personnel that mould the perception towards health care centres. Employees expect his company's hospital to understand the problem and consider providing treatment admission provisionally.

The above description shows the benefit of health care facilities for one Hadi household because he was employed with the Steel plant. However, most of them employed with Chas Nagarpalika remain without any health care guarantee. The socio-economic background of the patients impacts the health care workers more than the employed or unemployed status of the patients. It is also significant to note that the lack of awareness among Hadis about the means and processes to utilise available public health services is also a matter of concern. This lack of awareness is also reflected at primary health care centres and Chas sub-divisional hospitals. A few Hadis do not want to go to government hospitals for illness. They are not interested to know about the hospital. Sanjay Hadi, 22 years old, said, "I am not aware of Chas Hospital, nor have I visited it. It is somewhere on the outskirts of Chas. I was going to the Block office with a friend of mine when I learnt that the hospital is in the same direction on Purulia Road. As of now, I manage consulting doctor Ganesh, who practices in our neighbourhood." It is observed that Chas Hospital is visited in cases of major illness when doctors Vimal Singh and Ganesh are unable to cure despite treatment. At times, both the doctors, Vimal Singh and Ganesh themselves refer patients to Chas hospital which can help in saving money and perhaps proper diagnosis.

For two reasons, entry to a hospital during illness appears intimidating to most Hadis. Firstly, they are not knowledgeable about government hospitals' procedures and facilities/ services. Secondly, at times, the very set-up and behaviour of health personnel intimidate them. Experiences of disrespect and ill behaviour by staff at the hospital during entry and exit are constant. The same impression is conveyed back to the community.

Samina Devi, a widow and wife of Jogan Hadi, recalls visiting Chas Hospital to consult with her husband. I do not know the illness. She said, "*Bus nae janeio kintu nae bachlo. Dactarvan nae kuthcu kare paralkae hau. Kaho halthin ke khonki (TB) ke beemari bhae gael halae okra. Yahan kaise hovon hai aisan beemari ke jeeye le le hai* (I do not know more but that my husband died. Doctors could not save his life. We were told that he had severe tuberculosis when we consulted. I do not understand what kind of illness occurs here that it takes one's life)." Samina Devi lives with her two daughters and one son. The eldest is Kanhain, her son, who is married to Bhado Devi. He once narrated, "My father had developed severe cough at this last stage. I do not know what happened, and we took him to a hospital near Check Post, where they suggested visiting Chas Hospital. It was my first visit to a government hospital. The laboratory staff was nice and helpful in informing me about the illness. After consultation with the doctor, certain investigations were suggested in the hospital. It took 1 week to conclude that my father was diagnosed with TB. We were shocked. Certain medicines were given free of cost."

He went to the investigation department, where one person informed her about the disease and how to take medicines. It was a trauma to find a life-threatening disease in his father's body. He had some information that people do not survive, but the same person in the investigation department said not to worry. The medicine will cure him. But his father did not recover properly. His mother has little faith in medicine prescribed in government hospitals. Samina gets emotional to recall the loss of her husband so early. She is now 50 years old. Both mother and son described the experiences of visiting Chas hospital slightly differently.

Samina Devi does not want to visit again. But Kanhain said, "Maybe the disease was very chronic, and we visited quite late for a consultation with the doctor." In the same house, mother and son have two different perceptions about the hospital in the similar health crisis they faced. The above factors appeared to have strongly moulded the thought process of many Hadis about government hospitals and doctors. Hence, the notion that a government hospital will provide free health care is no longer significant to their life and preferences in seeking care. The loss of a household member with chronic illness has generated and shaped the perceptions of Hadis in illness. Patients mostly relate the causes, if they succumbed to illness, to the hospital, doctor, and poor financial conditions but not to other reasons. During my fieldwork, some of the doctors also reminded me to think about why a few Hadi died due to tuberculosis when there are good and effective medicines available in government hospitals. Hadis are perceived to be unaware of certain diseases and the threat associated with those diseases. For Hadis, the considerations are very different by nature.

Pari Devi, at one point, says, "*Toen kakro se pooch le, sabhe parivar yehe kahthoon ke kahil jeebein hospitalva, davae-daaru milbain karo vaha.* (You ask anyone, most households will tell you that there is no benefit in visiting the hospital (PHC); when I do not get any medicine, what is the use!)" Further, she says, "I could tolerate abuses (verbal), if at all they provided medicine. At best, I could save some of my household money on health expenses. If you think I am lying, please go to the hospital and cross-check with the doctors. After delivering my grandson, I took him to a private doctor at Chas bus stand. The doctor behaved well with us, but his compounder was not nice. He would behave as if we were there to beg for something. We also paid the fees of Rs. 100. That fellow compounder had not valued our faces at all. If we do not dress well with new clothes, it does not mean that others will misbehave with us. We have so many basic needs to fulfil. If our dress is a little unclean, it does not mean we like dressing like that. We spend money on detergents and soap, but there is no water to take a bath every day at home. Everyone has their household problems, but do we beg and seek sympathy? Their work is to diagnose and treat the patient, that's it."

In the above narrative, a woman does show annoyance with the assisting staff at a private clinic. She differentiates the behaviour of a doctor and an assisting person. When an infant falls ill, a doctor would normally behave nicely, which is received as a good gesture. Pari Devi compares the behaviour of a well-behaved doctor who treated her grandson and a compounder who did not behave nicely. She says, "When I asked the compounder about the doctor's consultation timings, she told me the date of the next visit written in the prescription. He said why did you not ask the doctor? Moreover, don't you see the dates written in the prescription?" This story helps to show the behavioural aspects of a compounder about the physician. It is also possible that the Compounder was unhappy with his work or that his employer, the doctor, would have scolded him, or that he was not paying enough money, but his work demands longer work hours. There may be many more reasons that can be assumed. But what is quite important is that Pari compares and finds the compounder's behaviour more disturbing, possibly for one reason: that the doctor behaved well but not the compounder.

In most of the interviews, it was noticed that a majority of Hadis have a similar perception of public and private health care services. Yet, perception varies with each household's experiences. Thus, behavioural discrimination is reflected in seeking health care at health centres in the preceding discussion.

During discussions with the people at Hadi Cooli, it was found that there is a deep fear in the minds of Hadis while visiting government hospitals. The choice of health care centres is not only dependent on the costs but also the treatment provided during illnesses, illness cured, and the type of social interaction between patients and doctors in the clinical set-up is crucial. There is a process of social construction in a patient's mind about the medical system and medicine as a whole. Once reported, the efficiency and capacity to diagnose and cure an illness takes a certain period for diagnosis and treatment. The social construction of medicine also depends upon patients' experiences in the process of availing and accessing treatment from health care centres. Health care centres comprise doctors, assisting staff, medical equipment, the available modern support system, medicines, etc. In this context, it is pertinent to understand the ideas put forward by Rosenberg that the construction of disease is an important aspect. However, he uses another concept to explain the construction and calls it a 'frame'. His attempt to avoid the term social construction "Tended to overemphasize functionalist ends and the degree of arbitrariness inherent in the negotiations that result in accepted disease pictures....chosen instead to use the less programmatically charged metaphor 'frame' rather than 'construct' to describe the fashioning of explanatory schemes for particular diseases" (Rosenberg 1989, p. 4).

There is a stage when a patient begins to frame about the disease and its intensity and probable harm. To illustrate the meaning of a disease, it needs to be deeply associated with an illness and experiences at health care services. The description about a, "Disease begins with perceived and often physically manifest symptoms" (Rosenberg 1989, p. 5), that may not be understood if not fallen ill. A patient with manifested symptoms reports to a physician and, in turn, learns the process of giving certain meanings and names to the disease he/she is suffering from. In the above account, the disease gets constructed when an illness is reported and a health-seeking process occurs. Rosenberg (1989) believed that disease can be interpreted as the direction of working out in a pathological process at the first level. Those suffering discomfort construct ailments and negotiate attitudinal responses, but non-human beings (animals) cannot construct the diseases but also suffer problems. The notion of a disease is understood when it is described. Hadis could not relate the problem of TB at a time when several of them were found to be affected by it. They could only understand the intensity of a disease when reported at the hospital and the disease was diagnosed. However, they could not be treated in some cases due to delays in reporting.

Therefore, their experiences during such illness are based upon the kind of treatment availed. Rosenberg writes that any disease does not exist in our culture as a phenomenon until the society agrees to it. The perceptions on any illness and disease gradually get constructed that also include the role of physicians and laypersons. Coughing was not considered as a threatening disease and Hadis would not report it to a physician until it affected their everyday work. TB was, at initial stage, misunderstood as normal cough and cold that occur in most persons.

In the above discussion and interaction of Nuniya, as discussed in the preceding section of 'doctor-patients', it appears that Nuniya Devi has not found satisfaction in the doctors' chamber. She wanted to explain her illness in detail and tell the doctor how the illness creates unease in her routine. She could also not explain in detail and asked the doctor to tell her what happened. Her dissatisfaction was with his not telling her what she suffered from and also being asked to leave. An explanation-seeking behaviour is found among the Hadi patients. The fulfilment of this part of the interaction plays a deciding role. She also felt that without listening, how could he conclude her illness? The interaction also points out that Nuniya Devi is an old woman and should have been attended to. Her illness may be minor and common among old persons, but her expectations from the doctor appear to be high. The doctor must have understood the symptoms and prescribed the necessary medicine. The treatment was provided, but the narrative shows that the patient found this experience with the doctor unsatisfactory. She believes that without listening properly, a treatment cannot be performed. These perceptions explain the hierarchical relationship between the patient and the doctor.

This results in the Hadis using their coping mechanism for initial and minor illnesses, which is reflected in the following description. Lamti Devi was found to be buying medicines at Chas Medical Store, the nearest and most familiar shop for Hadis. She says, "Babu (dear), I am not keeping well. For the last 2–3 days, I have been unable to sleep through the night. What has happened? How do I explain? I get continuous headaches, and it appears to be a disease called half headache, i.e. migraine (migraine is colloquially referred to as *adh-kapaari*). Usually, the beginning of the evening brings increasing headaches. But you would perhaps better understand what this disease is! Sometimes a lower portion of my leg also pains. The pain attacks in intervals. God knows this will be cured or not!"

Lamti Devi is a widow and aged 52. She lives with her married daughter and son-in-law. She kept her daughter in her own house after marriage because she did not have any sons. Romila is Lamti's youngest daughter, and she agreed to live at her natal home. Her father, Paltu Hadi, had passed away, and this was the reason for her decision to stay at Hadi Cooli. Lamti Devi got five paracetamol tablets from the chemist. He told her to take half a tablet twice daily after breakfast and dinner respectively. She told me that this headache is quite an old problem. When I suggested to her, "Why don't you consult and meet a good doctor?" She found it surprising and asked me how she could go to a good doctor. Who will pay for the fees and costly medicines? She narrates, "No dear, the doctor's expenses are high these days. I told you about my illness, but I hope you do not mean that I will visit a bad doctor (*baed* refer to a big doctor). Now, I do not earn. I live with my married daughter and son-in-law, and it does not always look nice to ask for such expenses. By the way, my son-in-law is a nice person; he gets medicines when I need them. I have three grandchildren. I do take care of most of the household chores. My daughter also works in two houses. She adds to the household income. At times, she gets food from her employer. The poor learn to live with minimum earnings. The doctor's expenses have increased steeply."

I suggested that her health is also very important to her daughter and the son-in-law, and they would be happy to see her healthy and that she should think from their point of view. This comment makes her happy about her household members loving her, but she again comes to the same point as she narrates, "The cost of living is very high these days, dear. Children's expenses, their clothes, school, and so many other unwanted needs arise every month. I have two granddaughters, and we have to plan for their marriage in a couple of years from now. Her parents will manage the marriage, but after all, I should also help; they are my granddaughters. I do not have good savings, but I have one or two pieces of jewellery I can give when needed. For these reasons, I try to minimise my expenses on doctors. Just two to three tablets keep me good and running. If you eat good food, then you will be healthy. At times, I feel like having a tasty and spicy curry. Sometimes, my son-in-law brings chicken from the market, and I would make it tasty with a lot of gravy. They also savour it. Let me tell you it is a fact that if you eat good food, you will not make rounds to hospitals. Then you will also feel that our doctor in Mohalla is perfectly fine for us."

There are several benefits to this interaction process at davaee dukaan. While the medical shop owner earns good money daily, Hadis must not wait at the shop for more than 5 min as patients. They find it very convenient to buy tablets, even for Rs. 5, which is never possible at a doctor's or private clinic. The cost of seeking health care increases depending upon the type of care sought. For instance, the local or neighbourhood doctor charges Rs. 30–100 only. In addition, both the doctors, Vimal Singh and Ganesh provide some medicines if the patient requests and to those who are very poor. Mostly they have some medicines given as samples by the medical representatives.

Doctor Vimal Singh says, "I do not charge on sample medicines. For those patients who are unable to pay and request some free medicine, can I provide it. But it depends on whether I have a medicine stock in my drawer. Second, giving medicines to the needy creates a good rapport with those households. You know there is a growing competition among doctors in the neighbourhood." He explained that there is only one physician named Dr. A.K. Singh, and the rest are like him (RMPs) and do not have any degree or proper certificates. He is somehow considerate towards patients, and they belong to various social groups, namely Bauri caste, Hadi caste, Shakaro caste, and a few Muslims as well. He claims to be one of the oldest practitioners in the locality, and his patients have faith in his treatment. Doctor Vimal Singh says, "Sir, belief is precious; I have secured it after years of practice. Of course, the practice has gone down due to the increased number of practitioners. However, all of us have our localities to practice. You know my son has completed his M.Sc., and he is about to complete his B.Ed. degree soon. After getting B.Ed. degree, Rajesh should get a teaching job in some school in the state of Jharkhand. Then gradually I can leave my practice. The struggle will be less. I will take some rest; after all I have been working since my childhood."

The experiences of Doctor Vimal Singh highlight the fact that practitioners also struggle for patients. The growing number of RMPs has reduced the number of patients per doctor. But the medical shop owner has a different narrative. The number of patients has increased, but with the increasing awareness and income level, many patients are now visiting trained physicians and specialists. However, the Hadis still depend on local doctors and medical shops for treatment. After all, it is the cost of medicine and doctors and investigation fees that matter. Most of the physicians and specialists will definitely suggest certain investigations and make extra profits. This cruel reality does not allow Hadis to go out for treatment unless it is serious.

In addition to the doctors in the localities, the medical shops also provide curative services as healers to the Hadis and many others. This minimises their expenses on health. The absence of a prescription and not paying fees to the doctor is only possible at the medical store. The power of knowledge to cure minor illnesses is asserted by Suhail, who owns a medical shop, "I can give better treatment than both the doctors (RMPs) who practice at Hadi and Bauri Cooli. The only difference is that I do not have a clinical set-up, and I cannot prescribe medicines on prescription. But I give the right medicine to the patients who visit directly. Generic medicines are low-cost with the same effectiveness. Sometimes, RMPs in the locality prescribe duplicate medicines that have lower effectiveness on the body. These medicines are locally made in cities like Ranchi and Kolkata cities. In such medicines, the quantity of actual medicine is very low compared to those from big drug companies. Patients pay less for duplicate medicines, and the duration of treatment is longer due to their low effectiveness. The prescriptions are written while mixing both duplicate and original medicines."

When original medicines are not prescribed, there is a delay in getting cured. In this way, firstly, the total cost of medicines, including treatment, is higher. The duration of taking medicine increases. The low level of effectiveness of those medicines prolongs the treatment. Second, there is a bigger margin for these medicines in retail shops. Hence, when a doctor prescribes these medicines, he also gets a certain commission from the medical shop. On the pretext that I should not reveal this fact, Suhail says "Doctor Vimal Singh has now started writing those low-cost medicines but also less effective medicines and gets a commission. Earlier, he used to send all patients to my shop, but now the number has been reduced to nil. He refers to another medical shop where medicines of good companies are fewer. After all, I am also in this profession for the last 20 years." Whether it is true or not, business rivalry cannot be verified given Dr. Vimal Singh's popularity. In this description, an important aspect emerges in medical practices. There is a possibility of a nexus between medical shop owners and doctors who are obliged by the pharmaceutical companies to prescribe their products. But quite significantly, what Suhail is trying to inform is that there are medicines in the name of generic section, but those medicines are locally made; rather, they are duplicate medicines and referred to as 'D'. Some of these companies put less quantity of medicine composition than the printed and prescribed levels. Such practice compromises the quality and effectiveness of medicines, and companies, along with all the stakeholders, maximise benefits.

Whether it is RMPs or medical shop owners, all of them have good income out of the health care services sector in the area. Of the total income, almost one-third of Suhail's business is based on the medicines bought by patients with prescriptions and those who seek medicines directly from him. However, there is a large turnout of patients for Doctor AK Singh. He prescribes good medicines, and his patients come from far away, from faraway localities, Mohallas, and from across social groups. The prescribed standard and good medicines mutually help the doctors and Suhail businesses.

Dr. A. K. Singh does not cater much to the medical representatives and asserts in an interview, "I already have a permanent job; I get a handsome salary. I do not need anything extra in my practice. Thankfully, I have a good number of patients, and that is sufficient. I pay three thousand rupees as rent to the owner of my clinic. There will be some patients who will seek a concession on my fees, though it is only 100 rupees. But Hadis generally do not visit my clinic. They rely more on other doctors who speak their language and belong to low castes. I do not go for home visits like a few RMPs who frequent their houses. However, I do get a good number of patients at my clinic every day." His honest attitude not only helps him retain his old patients and households but also increases the credibility of the medical shop.

In the preceding discussion, certain findings revealed that doctors have a dual role in providing health care to Hadis. At one point, they appear like a messiah and protector of sick and diseased Hadis. To keep a good rapport, doctors behave politely and give time to talk and listen during the consultation and outside the clinic, if found somewhere, provide certain medicines at low cost or for free to the needy. Second, they also make money through commissions from medical shops and get gifts from medical representatives representing drug companies.

In total, Hadis are at the receiving end on most occasions seven when lower-quality medicines are prescribed, as discussed earlier. Hadi patients are not aware of the aspects of the mutual understanding a doctor and a shop owner have and the benefits given by drug companies. The social relationship between patients and the doctor is deeply rooted and is more than merely that of a doctor and patients' relationships in the locality. There are certain aspects of trust and mutual respect. But this relationship does not stop the doctors from benefiting from drug companies. Those who do not want to pay doctor's fees or cannot afford doctor's fees directly visit the medical shops. Sometimes, the medical shopkeepers suggest that the patients visit a doctor since sending patients to the doctor will add to the number of patients at either Dr. Vimal Singh or Dr. A K Singh's clinic. The Medical shop person knows that after meeting the doctor, the patients will return to him. The number of medicines sold will increase with the increased number of prescriptions.

In June 2011, I had a good discussion with Jhomdai Devi, 47 at Doctor Vimal Singh's clinic. She said, "*Hamni davae-daaru par kharcha nae kare parbin. Badi giraani ho. Kakhnau thik nae lago hai kintu sabhe chot-mot dikkatvo mien dactor thin kii jaebiin.* (We do not spend much on treatment or medicines. It is very costly these days. How can I go to a doctor for a minor illness?)" She further narrates, "Take my example my son Kamal asked me to see the doctor and gave me a hundred rupees in hand. There is severe pain in my ankles and the lower area of my nails on both feet. How do I explain to you that my life is full of pain? Initially, I took some tablets during intervals, but it gave temporary relief. The pain again resurges after weeks. Yesterday the doctor gave me some tablets, but they did not work. Today, he gave an injection and also prescribed two medicines." She took out the prescription and pointed towards me, saying, "See this paper, you would know what is written; after all, you are also educated."

The doctor (RMPs) says that in the district of Bokaro, hospitals and doctors have not been able to cater to the health care services of the poor. There will always be some news that the Civil Surgeon would take action against RMPs, but before taking any action, the question is why there is a lack of better health care services in the town. The services of RMPs are not welcomed but patients have high faith in RMPs who are providing immediate treatment at the Mohallas and homes.

Recalling his injured leg, due to a road accident, Ashish went to the doctor (RMP) in his locality. At times of sudden exigencies, getting immediate health care is more important than bargaining about the cost of treatment. Ashish says that the total expenses calculated at the end of the treatment were higher than what he may have paid to *baed dactor* (a famous doctor/physician). But the comfort level with the doctor in the locality is the main reason for visiting the local doctor. The distance to a *baed dactor* would cause greater pain and discomfort, especially with an injured leg.

However, the charge for most of the trained doctors is not more than 250 rupees per visit. But for the consecutive visit, the patient has to pay a consultation fee again. In case the visit is made within 1 week, there is a facility to waive off the consultation fee. However, the cost of getting X-rays, investigations, and laboratory charges exceeds the consultation fees of any doctor. It is common talk in the locality that a fair amount is paid as a commission to the referral doctor at a fixed rate for each test by the laboratory. The total earnings from such commissions depend on the number of patients sent for laboratory tests. This entire process of passing through a chain of qualified practitioners for medical help influences the patients' perceptions. Such processes gone through with health care providers at times influence both the illness perception or behaviour and the health care sought. Nagraj Hadi shared that private hospitals are costly because they not only give medicines but also recommend further investigations. Illness perceptions are a kind of cognitive belief a patient has about his or her medical conditions and discomfort felt in the body (Katavic et al. 2015). Mechanic (1986) observes that illness behaviour is a state of the human body that is shaped by socio-cultural and socio-psychological factors in various forms. In other words, the inability to do everyday work and the pain in hindering daily routine activities refer to states that seek the attention of the person. It was found during household interviews that Hadi elders and their youth also adopt a delay mechanism in seeking health care. It is seen that, by and large the level of morbidity among Hadis is high due to this delay in seeking treatment. Some data is suggesting the existence of infectious diseases among Hadis too. Two persons were diagnosed with Tuberculosis and put under medication. Jhagu Hadi, husband of Simar Devi, and Jogan Hadi, husband of Samina Devi, were two patients suffering from TB, as discussed earlier. TB caused two other deaths.

Interactions between health seekers and providers do not happen in isolation but they form a significant component of the given social structure. The interaction between Hadi patients and health care staff at government hospitals does not satisfy the care sought. There is a widespread notion among Hadi Cooli members that there is free treatment at government hospitals, i.e. patients are entitled to get free consultation, all medicines at no cost and medical investigations shall be done without any charges. However, in reality, the hospital has a different set-up and story. None of the government hospitals are equipped with medicine counters except for the national vaccination programmes. Even the family planning programmes are conducted at Chas Sub Divisional Hospital or BGH, but the surgical procedures are not available at the Chas Sub Divisional Hospital. All such patients are referred to Bokaro General Hospital under the government scheme of family planning. The efforts of reaching out to a primary health centre are made by Hadis, but as seen in some of the narrated instances, the hierarchical social structure within the PHC reflects the lower utilisation of health care services. There is a regular trajectory of operating in Hadis's life operating. Some of them visit PHC and then turn to other doctors if they are not satisfied with the treatment and experiences while seeking treatment.

First of all, the data shows that attempts are often being made to make use of government health care, but the health care providers acted as one of the barriers to Hadis. The situation should have been reversed. Second, the non-availability of doctors on time at PHC also discourages the patients from visiting. In continuation to this, the behaviour of health personnel at the counter, i.e. the individual who provides slips to meet the doctor, is also found strange by Hadis.

A patient has to stand in a queue at the reception counter and request for a consultation slip to be given in his or her name, address, and illness they are suffering from. After getting the above-mentioned information, the person at the counter issues a slip that is to be used for consultation. The first interaction takes place at the slip issuing counter that in a way reinforces the caste and community background of the patient leading towards a discriminatory behaviour. Bablu Hadi recalls such an instance during his visit to the hospital along with his father, Raju Hadi. His father requested an early consultation to meet the doctor and asked for the slip at the counter. He was scolded, "You all (Hadis) come late to the hospital and want to become VIP. It has become a habit for most Hadis. Go and wait in the gallery for your turn." The patient's narration in the hospital gave one instance of ill behaviour towards Hadis. Bablu felt humiliated because his father was unwell and was having pain in his stomach for a long time. Later, he was diagnosed with cirrhosis in another hospital. After the above experience at the counter for taking a consultation slip and waiting in the queue is perceived in a negative light by Bablu. In the above step, finally, when the patient meets the doctor, the doctor turns a deaf ear to the illness description. This visit at PHC is in the mind of Bablu even after a gap of 5 years or more. Past experiences remained in his mind, and consequently, he does not want to visit any illnesses to the government hospital. The illness experiences of his father's disease have made profound impact on his perception.

Ganesh Hadi's daughter's case also corroborates the view that Chas Hospital is not accessible to them. The experiences of visiting a hospital or clinic in the past appear as an important factor that influences the mindset of Hadis in society. There are other factors, such as ecological forces, that determine the pattern of health problems.

The time given to the patients, the manner in which the doctor speaks to the patients, and listening to the patients themselves are skills which most doctors miss. There is a perception among the doctors that certain symptoms make them clear about the disease, which a patient takes a long time to explain. However, they do not want to waste time to prescribe medicines. This is observed in the process of health-seeking behaviour of Nuniya Devi at Chas PHC. Her narrative is discussed earlier in this sub-section, which talks about how the doctor does not provide time to listen to her illness. Doctors are experienced; they have an understanding of the symptoms of illnesses. But for a patient, in this case, Nuniya did not find it helpful, and her satisfaction level was quite low when she did not get to narrate her illness to the physician. These kinds of experiences are shared to the household members and friends in the community. The construction of perceptions not only takes place in the process of seeking health care services for an individual's illness, but it is also shaped when others discuss their experiences.

There is another problem that exaggerates the level of illness and suffering due to the above reason, the reason of mutually shared experiences. Hadis, at several points in time, did not return for follow-up checks and consultations that were mostly advised to them. But the lack of trust, the feeling of not being given importance, ill-treatment, and finding no improvement in their health construct a mindset that restricts many Hadis from going for follow-up consultations. In these situations, illness increases. As mentioned above Nuniya believed, "without listening properly, a treatment cannot be performed." These perceptions explain the hierarchical relationship between the patient and the doctor, but it is a story quite common in most doctor-patient relationships across classes in India. It dissuades people from seeking free health care provision and is too poor to pay for qualified practitioner care.

One such example is Raju Hadi's, 38 years, case. He suffered from liver damage could not understand the seriousness of the issue, and was dependent for too long on Doctor Vimal Singh. On another occasion, I visited his clinic, and later the doctor explained his illness to me. The doctor admits that he never knew that Raju has developed cirrhosis due to alcohol consumption. In such illnesses, generally, he sends a sample for investigation, but in this case, he had no idea that Raju had already developed that is cirrhosis, a disease linked to the failure of the liver. It was a first case of liver failure in the experience of Dr. Vimal Singh. The doctor told me that neither he has the expertise to provide treatment in such diseases nor could he refer without putting in his efforts.

However, there is no established scientific relationship between water and liver infection. Alcohol dependence (*daaru*) is also one of the reasons for liver troubles. Ramesh Hadi, aged 69, explained other dimensions that Hadis die at an early age as they are addicted to consuming alcohol. Too much alcohol consumption creates several diseases in the body, and one really does not know about its full impact on the body. It is believed that there are chances of longevity among the present generation of Hadis due to the declining consumption of alcohol alcohol consumption. The discussion on alcohol took place when I was in a meeting with youth in community hall and learnt that alcohol dependency increases burden on households, women quarrel when in financial crisis, etc.

Raju died due to alcohol addiction in his 40s. His wife, Pari Devi, lives with her only son. Her husband could not see his grandson, so he had a lot of plans. He used to say that grandchildren would make him happy in his old age. Pari Devi's eyes were full of tears as she narrated this, but she kept sharing her memories of Raju Hadi. Raju Hadi was never careful about drinking. Every afternoon, he would drink while returning from work. He would earn rupees Rs. 4500 as salary from Nagarpalika apart from earning extra income from part-time jobs. His wife spent roughly Rs. 10,000 rupees during the last few months of his treatment. The household was already indebted due to the visit to Ranchi for his treatment. They had to pay an interest for rupees 15,000 to the Sahu (money lender) at *Sahu Bazar* on Chas main road. The interest payment was Rs. 200 rupees/month for 2 years, but the principal amount has not gone down. She is not aware of the rate of interest and the amount that needs to be paid. It is important to understand her financial condition, which is precisely why those medical expenses rose to a high level, and now she finds it difficult to meet those. Hadis suffer severe illnesses due to low-income level and coping with such severe illnesses with high expenditure at private health care is difficult. Initially, the local doctor gave regular medicines for indigestion and acidity. Two types of syrup were to be taken before food in the morning and after food in the night every day. It was a kind of enzyme syrup that helps all minor disturbances in the liver related to acidity and indigestion.

Raju was also given Ayurvedic treatment. He visited the Ayurvedic doctor in the main market of Chas. The doctor encouraged him that nothing drastic had happened and that by controlling his food intake, it should get better. Several suggestions were given about food habits. He was asked to stop eating spicy food and control his meat consumption. One month of medicines would cost nearly Rs. 500 rupees. The first month of treatment did help his stomach trouble. The powdered medicine suited his indigestion and the pain in his stomach. The treatment took 2–3 months, but every month, Raju visited the Ayurvedic doctor, and he was asked to wait and take medicine properly. Raju's anxiety would rise when the doctor did not answer the question about him being cured of his illness. An Ayurvedic doctor does not explain the causes of disease as he felt. His doctor does not ask for some investigations. Rather, he prescribes to stay on his advice about food to be taken and must be avoided. Ayurvedic doctors have many reasons for the disease, unlike allopathic doctors, who rely on medical investigations. Pari Devi had to struggle during this period because the Ayurvedic doctor suggested Raju stop drinking alcohol during treatment, but he did not take the doctor seriously. The medicine reduced his pain and indigestion, but it would not be effective for a long period until proper food avoidance is not maintained, as told by the Ayurvedic doctor. Nausea did not reduce. Even after a small meal, he would feel uncomfortable. The doctor was helpful, but every month, when both of them visited the Ayurvedic doctor, he would tell him to take the medicine for one more month, and hopefully, it would cure him. Pari felt that "Even the Ayurvedic doctor was not sure about his liver failure," but he was cautious about the health of Raju. Raju's face was turning yellow due to the disease.

Pari narrates, "But he never stopped drinking alcohol. The doctor whom he visited at main road suggested him every time to abandon drinking. After a short interval of non-drinking, he would start again. Then, after 5 months or so, we left that doctor and started consulting Ayurveda doctor. The treatment went on for almost 5 months. Every month, five to seven hundred rupees would be spent on medicines and fees. It wasn't easy to meet the growing cost of treatment, and the doctor kept telling us to wait for another month and that he would be fine. Hence, we stopped meeting the Ayurvedic doctor after a few months. One of my relatives asked me to visit him in his hometown, Dhanbad. After reaching, he handed over some leaves which he had got from his neighbour at Dhanbad. The grounded leaf was to be administered with warm water. This medicine was for 3 days but did not help reduce Raju's disease. Perhaps fate was written differently because no medicine could cure his illnesses."

The data illustrates more than one dimension of treatment sought when Raju fell ill. The liver failure became a major disease, but it was not noticed in the early phase of treatments. Stomach pain and nausea, etc., are quite commonly reported illnesses. People do not find it life-threatening; more over Raju and his household members did not assume the severity of troubles in the liver.

Doctor Vimal Singh said, "Vomiting and fever are the symptoms of many other diseases as well; hence, I manage such patients and their illnesses with antibiotics and a few other tablets. But the case of Raju was so complicated that I could not even imagine that he had a liver failure. Without knowing the disease, I tried my best to treat and save him, but what could be done, I put in all my efforts; after all, life is not in our hands, but this case opened up my mind for future treatment with similar symptoms. Actually, there is a high tendency of alcohol consumption among Bauri and Hadi castes; they drink beyond their capacity and lose consciousness. I do not know who would change their drinking habit but it ultimately affects their health."

In this description, it is important to note that the doctor was unaware of Raju's disease. It is not his ignorance but his perception that this patient was a normal patient who would have a minor illness like stomach upset, indigestion, and related illness. The medication was prescribed symptomatically to subside the discomfort and regular pain. It was much after the diagnosis that Doctor Vimal Singh got to know about Raju's deteriorated health conditions. He did not refer him for medical investigations he sometimes prescribes for others.

The above narrative of the death of Raju Hadi brings the matter of health care into discussion at the household level. Gradually, his health was deteriorating, but for quite sometimes he continued working in Nagarpalika. He was also able to eat food, and at times, he would ask his wife, Pari Devi, to cook a bit of spicy food. His wife cooked simple food without any spices for Raju for 1 year. The entire day, he would be fine, but after dinner, Raju would feel some pain in his stomach. The syrup given by Doctor Vimal Singh generally worked before sleep. The spouse hardly had any doubt about the treatment and experience of the doctor. His death was blamed on his drinking and not listening to the pleas of his wife, Pari Devi, to stop drinking. Raju himself was responsible for his deteriorating health.

In other words, Pari was always worried due to his heavy drinking habits. She never liked his stubborn behaviour. She says, "I repeatedly requested him to leave alcohol; you are already sick and unwell." His health began declining 2–3 years before he passed away. While going to work in the morning, he would drink *daaru* (alcohol) at Bauri Cooli and then go to work. One good habit was that he never fought or was involved in manhandling during work, thank God! Gradually, with every passing day, his health started deteriorating, and we thought that it was only due to his indigestion and he would recover soon. With some medicines from Doctor Vimal Singh, he will be alright."

She continued, "Every month, we had to pay 500–1000 rupees to the doctor for his regular medicine. The day he (Raju) would eat meat at home or any feast in the community, he would not digest properly and often would have loose motions and consequently dehydration. I had a lot of difficulties with his addiction, but he fulfilled our basic needs of food and clothing. He always managed to work. One day during the summer, he got bedridden. I also promised to sacrifice an animal or bird to Mansa Devi (local Goddess) to get my husband well; for a moment, I thought he might not survive. We took him to Muskan hospital in an auto immediately and admitted him in an emergency. Almost Rs. 3000 rupees were spent on that day for his treatment apart from the medicines and investigations done later. He was admitted for 4 days and then discharged. When we asked the doctor on duty, we were told that Raju had no serious illness. He should take medicine for 3 months, and that should cure his illness. But 3 years passed, and he did not recover fully; he had the same problems until he passed away."

While managing Raju's ill health, Pari Devi learnt different aspects of life. Doctor Vimal Singh never revealed and explained the result of such an illness. He always gave a positive explanation, and moreover, he empathised with the patient and Pari Devi. Perhaps Raju, who kept doing his job at Nagarpalika, never thought that his liver was damaged seriously. His inability to stop drinking was not only due to his addiction but also due to his surroundings. Several men drink the locally made alcohol that is sold in Bauri Cooli, North of Hadi Cooli.

Raju passed away at an early age. He did not even cross the age of 40. Raju left behind a wife and three children. His son Bablu was not even 20 years old when he passed away. After the demise of his father, he had to start working. Nagarpalika provided him with a job with a monthly salary of 3000 rupees. In the coming year, his salary was supposed to be increased to 3500 rupees. This narrative substantiates the earlier explanation given by Ramesh Hadi that the longevity of Hadis was very low due to their drinking habit and low food intake. How Hadis themselves engage with the physicians and clinics during illnesses also shapes and reshapes their perception of illness. For instance, the field data of Raju incorporates that chronic illnesses made most Hadi families vulnerable to the process and procedures of treatment at an untrained doctor's clinic.

Pari Devi knew that the government hospital would not have saved her husband because its facilities were negligible; therefore, she took her husband to Muskan Hospital. The emergency ward at Chas Hospital was non-functional. In case of emergency, Hadis tend to visit Muskan Hospital, which is a private centre, and most of the patients in emergencies rush to Muskan Hospital.

It has also been observed that the Hadi households do not want Doctor Vimal Singh to refer them for certain medical/clinical procedures and those investigations that involve a process of testing and reporting. But the doctors explain the importance of 'evidence-based procedure', after which the investigations are also conducted under the reference of local doctors. It is seen that sometimes seasonal infections attack people in the area, like fever, cold cough, diarrhoea, etc., particularly in the post-monsoon months.

During the year 2010 and 2011, a kind of fever gradually attacked a large number of people in most of the Mohallas of Chas. It took the shape of an epidemic, especially in the poor households who could not take immediate treatment, and on average, most of the household members started falling to the onslaught of an unknown fever. Even during interviews with several other doctors, namely Alam and Jamal, who have a large number of patients in Muslim Mohallas (that are the Mohallas they cater to Sultan Nagar, Ansar Nagar, Ghaus Nagar, and Azad Nagar). There was a consensus among them to provide simple paracetamol with a combination of diclofenac and saline water to improve the patients' immunity. However, this epidemic was an unknown fever; a few people said that it was chikunguniya, but it was also not proven. The fever lasted 2–3 weeks as Doctor Vimal Singh narrated. He was busy swamped during this month, and everyone was scared. The local population shared that Muskan Hospital did not have even a single vacant bed. The fever was like a curse on the people of Chas in general and the poor in particular.

Doctor Vimal Singh finds it non-problematic to follow a Hadi patient's wishes and moods during their clinic visit. He says, "If patients do not want to go for investigation, we cannot force them. But when I have strong doubts, I ask the patients for them. Most of them want to get well in just two tablets and some medicines. How is it possible! Nowadays, the illness also arrives with full planning to drain the patient's blood and money. What can I do in such situations? Investigations indeed help us in giving accurate treatment. However, we first try medication on the basis of reported symptoms, and if it works then nothing better. You have to spend some amount when seeking treatment, but we do not charge hefty fees. These days' illnesses have increased. Everyone appears to be weak. I did not see such a trend until 10 years back, but now I observe there is a health breakdown among people of all age groups. How to tell you! I faced one incident in the year 2010. I began providing treatment to a Bauri person who was really ill. He had a fever, stomach disturbances and constant vomiting, but he was not ready to go for investigations, I suggested. I gave him antibiotic tablets and paracetamol along with a combination of Azithromycin and diclofenac. He appeared to be recovering initially, but suddenly, on the third day, his condition started deteriorating. We immediately rushed him to BGH emergency ward and admitted him in the private ward. I have good rapport among the households of Bauri Cooli, so a good number of patients visit my clinic. Vijay was admitted in the forenoon and he succumbed suddenly on the same night at BGH. He was almost 40. He has left behind three children and a wife. This incident gave rise to a new problem in my life.

The doctors at BGH told the household members that the investigation reports suggest that the death happened due to medicine reaction. Just after listening to this fact, all family members of the deceased started accusing me for Vijay's death. The family members planned to lodge a police case against me at Chas Police Station." The doctor's situation is also difficult, and one mistake puts them in difficult situations. It is never intended to harm a patient, but a doctor who is not aware of the various side effects of medicines could commit a possible life-threatening mistake. If one reads his whole narrative, it will help to understand the complex whole of the practitioner's view point based on his experiences.

The doctor continued, "But elder community members came forward and debated the matter. Finally, it was decided that I shall pay one lakh rupees to his wife. This compensation was heavy for me and it wasn't easy to arrange one lakh rupees. Then I persuaded some of the influential Bauri members in the community and told them to reduce the amount to rupees Rs. 50,000 otherwise I could not pay. After so much persuasion, they agreed to my request. This saved me from the unnecessary troubles of the police. How to tell you, this kind of incident happens in our life, and during practice, some get caught in such cases also."

There have been a few instances when patients were given the wrong medication, and it led to worse and a few casualties. The other fact is that despite a couple of such rare cases patients do not stop visiting the clinic. Doctor Vimal Singh gives an important point that it is generally crucial to treat Hadis. To them the illness begins when they are unable to cope at the household level. It is commonly found that illness has not been reported. Hadis find no problem in ignoring the feeling of being unwell.

A woman Romila Devi who is 30 years old says, "I could not understand how the labour pains began. I was preparing vegetable curry for dinner, and suddenly, I had to stop cooking due to pain. I sat down on the floor. In the next 1 h, I delivered a baby daughter at home only. My husband called Dai from Bauri Cooli and she handled the entire job. The next day, my husband informed the doctor and requested him to visit at home. The doctor administered one injection, and everything was all right. Now, I hear so many illnesses and diseases cropping up during childbirth. When you visit the hospital which disease and what will happen, you never know. I gave birth to all my children at home. It was very difficult during my last birth and I had severe pain for a week. However, it is not an illness, everyone feels this pain. The doctor was suggesting me that I should take some medication, but I did not want to spend unnecessary money. But the next week, my condition began failing, and suddenly, unprecedented bleeding started. My husband immediately rushed to the doctor and called him urgently. He administered two bottles of saline, and I was kept under medication for one full month, which I did not agree to initially."The need to seek medical support came at the last moment when Simar's health began deteriorating. The perception, and mostly a traditional Dai attends such at home birth of managing a delivery at home is quite common, and mostly traditional Dai attends such at-home birth.

In situations of urgent need, the doctor is readily available. This data shows a tradition of home birth in Hadi Cooli and also the illness behaviour that seeks medical support when the birth attendant fails to handle it. It could be illustrated in two ways; firstly, Hadis have been practising the tradition of home delivery. Secondly, medical intervention is sought only in emergencies, if there is a complication or delay in birth, bleeding beyond the normal condition, etc. Therefore, the traditional support system works, and simultaneously therefore, the traditional support system works and simultaneously, modern medicine's help is sought.

Immediate intervention over felt illnesses is of utmost concern and satisfaction. For example, the blood or urine examinations are not charged a handsome amount, unlike the private hospitals that do so in the city. Armstrong (1984) says that a doctor has to depend on investigation reports in order to diagnose the disease. However, the RMP doctors in the Hadi Cooli area are not bound by the investigation and reporting procedures due to certain reasons at Chas. There are two reasons, Doctor Ganesh says, "I think of referring patients for some tests but the patients generally do not adhere and refuse to do so. The cost of treatment indeed increases due to investigations. Patients pay just 50 rupees but I cannot tell you about the laboratory charges.

Most pathological labs in Bokaro charge differently. In addition to this, patients strongly feel that medical investigations are not necessary. In this situation what can I do? I prescribe and provide treatment on the basis of symptoms provided at the time of meeting patients. However, sometimes investigations become important due to hidden infections in the body. For all these reasons, it is not possible for us to practice in Mohallas and Cooli of Chas. After doing all treatment for Hadis and other caste groups, they do not want to pay for our fees." In this elaboration, other aspects of the treatment procedure and a feeling of lower earnings are reflected. The doctors do not necessarily prescribe to take investigations but ask when it is essentially required to treat patients. A few patients do not want to spend money on investigations and request the doctor to prescribe medicines without any investigation. There is some sort of clash between a patient and doctor and both would have different reasons, but solve it mutually.

It is a matter of fact that the number of patients for an RMP is always as many as at a trained doctor's clinic, whether investigation is prescribed or not. On average, 10–15 patients turn out at Doctor Vimal Singh's clinic daily, and this doubles in the post-monsoon season, when several infections spread among the people. In other words, an RMP doctor makes handsome amounts by taking a small fee. The fees vary from 20 to 100 rupees depending upon the capacity of the patients to pay. The nature of illness also governs the cost of a doctor's fee. These suggest an interesting feature and dimension of the doctor and patient relationship in Chas town. The perception of Hadis changes accordingly during the treatment procedure at the clinic and meeting the doctor. The close interaction process between both doctor and patient increases the level of trust and mutual understanding. Patients show increased reliability in the doctor and his diagnosis. However, Doctor Vimal Singh says that generally he would explain the problem and nature of treatment required after listening to a patient. This provides clarity in the minds of Hadis seeking health care.

Doctor Ganesh, on the question of investigations and what kind of commission returned to the doctor, responds about why he does not suggest investigation with a note of precaution: "Sir, the trained physician or doctors at the hospital get a certain amount of commission out of every investigation they prescribe and the patients would have no choice except to abide by the prescribed investigation. Even a few RMPs refer for tests and get a certain amount of commission from the pathology centres. But the huge commission is returned to the trained physicians monthly but not to the RMPs." RMPs also get certain cuts as commission from the number of pathological tests they refer to every month. The fact remains unknown to the Hadis. In turn, they tend to shower high respect and place trust in their doctors. On a couple of occasions during the meeting and interviews with doctors, it was confirmed that various pathological test centres give commissions to doctors in the Hadi locality, and it is accepted as a bonus among the doctors. I tried to discuss this with one laboratory at the city centre, but they did not disclose anything about the commission paid to the doctors in lieu of referring patients.

My point in bringing the narrative mentioned above into the discussion is to relate with the cultural aspects of health-seeking a relationship between the doctor and patient. The vital essence of illness and its experiences lies in the lives of Hadi and their health care providers/centres vis-a-vis accessibility to health care services. In this section, it is discussed that health and illness are the product of society's socio-cultural and political structure.

The above illustrations show that the experiences of Hadis with private doctors in the locality and in hospitals across the town have been shared among themselves. The shared experiences are important factor that guides them in future while seeking health care services.

In another aspect, as discussed earlier, it appears as normal when Hadi workers can pick up their work tools, namely Hoe (Kudaal), Weeding Hook (Khurpi), used for digging and weeding operations, i.e. Kudali and Spade (Belcha) and go to work every morning. These tools are used to collect and put the garbage into rickshaws. In the circumstances of their inability and discomfort due to minor illness, there is no need to either rest or visit a doctor. Quite often, the first stage of unease or bodily discomfort compels them to go to the druggist in the medical shop, and by this, Hadis could avoid and sometimes delay visiting a doctor. They find it easy and comfortable to narrate their problems and seek some medicines from the drug store person who explains its meaning and the usage of medicines in the case of each individual. Buying medicine in consultation with a medical shop owner or a chemist' without visiting a doctor is normal behaviour during non-recurring illness. For instance, one can have good or bad health but a recurring illness is not desirable.

The Hadi patients never felt acknowledged as equals with human dignity despite the humility shown by doctors and administrative staff during the visit they made to PHC and Chas Sub Divisional Hospital. The illustrations on both sides are important data to understand the behaviour at the hospitals. A few doctors opined that it is difficult to provide extra time to patients due to overburdened workload. In an interview, Doctor Ram, then district Civil Surgeon, differed and said, "There is no extra or overburden at the health centre. The doctors do not spend full hours of their duty. When a doctor is available for a short period, it is obvious to see the number of patients would be higher. It isn't easy to oversee the Department of Health. I have left a few months then will be free." There was observation that doctor is not staying on their duty period. The non-performing doctors and their inability to give proper consultation timings show the hospital in bad light and ill practices reflect on the patient's satisfaction.

An important point to discuss is how doctors' behaviour apparently reflects on the entire health care provisioning under government departments. The doctor responds and tells him to be polite during consultation hours and that patients have different opinions. The significance in the data is the kind of experiences patients bring back to their community and subsequent discussions; it also shapes and influences people's perceptions.

Lamti, a widow whose narrative is discussed in the preceding sections on the 'doctor-patient relationship' as well, gets angry with the kind of behaviour shown to her at the hospital. For this reason, the most suitable clinic is the one in their locality. For instance, Doctor Vimal Singh is always empathic during meetings at the clinic and provides his best, says Lamti. For these reasons, several respondents do not find it comfortable visiting government hospitals.

Many sociologists have studied hospitals and doctors' clinics Caudill (1956), Reader and Goss (1959), Apple (1960), Kleinman (1980, 1985, 1988), and have helped me to revisit the questions of patients and the impact of medical bio-technology and equipment on the health of the patients.

It is widely seen in the hospital administration and during the provisioning of health care services that the existence of conflict and problems in the hospital administration due to certain factors like domination of the doctors and technology settings, and therefore, it is not natural to observe certain problems mentioned above. In the process, patients are marginalised. The patients' existence is least important to seek health care services. The right to be informed about any investigation and the socio-cultural barrier in getting socially accessible is important to patients like Hadis.

3.8 Patients and Clinical Set-up During Illness

The discussion in the earlier section was about how doctors do not provide social space to the patients and do not give patient hearing on illness. This may be one of the reasons that Shah and Ahluwalia (1970) emphasise that sociology should be taught to medical graduates, to create the spirit to enable them to perform the social role of a Doctor. The Parsonian model of sick role is well established and its existence is found in the hospital and clinical settings, when a doctor and patient establish a relationship during treatment and the follow-up process. The patient's normative behaviour of sick role is defined and narrated across the social settings but the role of doctors is not defined. For instance, Hadis as a poor patient are expected to behave in a way that is submissive and non-questioning to the doctors, and other hospital staff. The behaviour of the hospital staff is to be accepted as normal even when they do not help with patients. The doctors' existence is a dominant phenomenon in the modern bio-medical medicine unlike the mutual relationship in RMPs clinical settings and with the other local healers in Hadi Cooli. It can be argued that V Sujata describes that those who have privileges mostly influence the bio-medical area and get chances to avail a better health care than the underprivileged classes (Sujatha 2003).

The presence of disease is considered unwell and dysfunctional in the life of Hadis as it creates hurdles in the daily activities of Hadis. For instance, when a Hadi accepts and obliges to accept the role of sick person due to some disease, he has to avoid work, may visit a doctor in a clinic, and follow up the suggested prescription. Most Hadis attempt to ignore visiting an allopathic doctor's clinic, but in serious cases, that deters them from attending work, and guides them to a doctor's consultations. There are cases when patients die due to illness and serious disease, but they do not accept themselves as ill. Raju Hadi never admitted that he was unwell and kept doing his work. But he also never stopped drinking liquor that damaged his liver. The perception of his ill health was considered somewhat non-threatening to his life. Sometimes a patient is examined to be quite normal due to his/her physical appearance but disease like cirrhosis existed in the body of Raju.

Till his last breath, Raju did not believe that he was suffering from cirrhosis. However, the Hadis are conscious of the fact that they are labelled as filthy, alcoholic, and unhygienic but do not dwell upon this labelling or attempt to alter it. With the advent of Western biomedicine, the labelling of an individual to a specific lifestyle disease became clearer. A process of diagnosis and treatment in clinical or hospital settings became more established. For instance, Tuberculosis (TB), for a certain span of time, became a nuisance to the lives of TB-infected patients in society. It could ruin the lives of an entire family and its dependents. Kumar and Mahapatro (2010) describe how social and economic costs of TB are very high for TB patients; it leads to several repercussions such as divorce, loss of job, shrinking of social relationships, and stigmatised life in society. But if the stigma of being filthy and unhygienic is added, the complexity of illness increases, as among the Hadis.

Anulup lost his father at an early age when he needed his guidance and support. His father died at the age of 40 due to Tuberculosis (TB). His father, Jhagu Hadi, was the only son of his father. Anulup had no paternal uncle. Naturally, the total responsibility of the household came to his shoulders. When Jhagu Hadi was in his 30s, he started facing the problem of breathing and coughing. Gradually it became a permanent illness for his entire life until his death. Household members could never realise that this disease will be the reason to Jhagu's death. They had less know-how. Indeed some sort of stigma that this *khonki-beemari* (TB due to its regular cough) is communicable and can be transferred to others Hadis somewhere knows that there are medicines to cure TB. There was neither Tuberculosis hospital in the Bokaro district, nor was proper treatment given at government hospitals. The hospital in Chas has the facility for medication, but perhaps the lack of follow up and absence of guidance to patients brought undesirable results. Dr. Dasgupta, who retired from BGH, said, "The number of TB cases was higher in the 1980s, and a good number of serious cases were referred from Dhanbad to BGH. At present, the reporting has reduced due to the availability of medicine in most hospitals, unlike two or three decades in the past."

To include small narrative on sufferings from the lives of Jhagu Hadi, his widow Simar Devi recalls the experience of her husband while he was suffering from Tuberculosis. Jhagu Hadi passed away due to TB, but his wife says, "One is not able to understand what is going on inside the body. The chest appears like it will burst or crack due to dryness, neither anything comes out from mouth nor can you cough up, you keep coughing like a weak dog keeps barking without reason at night."

BGH was earlier unavailable to the outside patients until it was referred and a guarantor as an employee of BSL put forward the request. In the case of Jhagu, his household members were not aware of the kind of suffering a TB patient would face. The lack of awareness and negligence increased his chronic conditions. It was much after when Jhagu started having severe cough, sometimes coughing blood and sleeplessness through the night, which forced him to visit the sub-divisional hospital of Chas. His mother narrates that 20 years back, the condition of Hadis was much worse in terms of poverty and lack of knowledge about the outside world and even about hospitals. The hospital staff and doctors would not talk and give proper guidance at the hospital, nor give attention during the illness. It is quite clear that Hadis never had proper access to government hospitals, and the social-physical accessibility was less, nor did they bother to comply with prescribed medication. The lack of regular doses in diseases like TB makes the drugs resistant and does not cure. It happened in the case of Jhagu, which perhaps led to his illness at the terminal stage, and Anulup lost his father. The illnesses and sufferings during his TB were unavoidable. For instance, Guddi Devi, wife of Vijay Hadi in the field, tells me that she feels vulnerable when she feels unwell. Nevertheless, she cannot express her illness to the household members nor go to hospital. The entire process of being sick or unwell disturbs her household. She calculates the expenses on treatment and medicines and knows that her household income is minimal. Until she is unable to cope with her illness and is incapable of attending to her household work, there is no need to visit a doctor and seek treatment.

When the above case of TB is considered and read together with Mckeown's (1979) thesis that the mortality rate in the nineteenth century declined not because of medicine, instead, socio-political factors played a catalytic role in reducing mortality due to improved housing, nutrition, potable drinking water, and sanitation. This theoretical position says Hadis could improve their health if their socio-economic conditions are improved, proper sanitation is provided, and clean drinking water is available. Nevertheless, the fact is that neither the socio-economic condition has improved nor has modern medicine reached the Hadis at an optimum level. Government hospitals are located far away. Hadis of Chas eat less nutritious food and have congested housing conditions, especially poor ventilation and lack of sanitation, and not merely due to lack of health care services in the vicinity. There is no denying the fact that access to health care services is crucial (Acharya 2010a, b) to a socially and economically weaker social group, which contributes to shaping the perception of Hadis towards illness. Baru et al. (2010) have examined the various dimensions of the question of health care accessibility. "Three major forms of inequities have been largely responsible for the persistent and even widening differentials in health outcomes: historical inequities, socio-economic inequities and inequities in provision and access to health services" (Baru et al. 2010, pp. 49–50).

Mathur and Sharma (1995) place illness and health in a social context in which not only the patient but also his whole environment and particularly those who try to heal him, find their place. Sigerist (1952) explained that being ill would mean to suffer and every culture defines its stand towards suffering. Guddi Devi considers suffering as a part of life of the poor. She lives in a joint household along with her mother-in-law, husband's brother wife, and children. It is a difficult task to manage a house of 11 members in just Rs. 10,500 rupees out of two earnings. The social situation in which an individual lives plays an important role in deciding the illness-seeking behaviour. Even in a situation when a person feels unwell, the household may discuss the nature of illness and discuss about the level of discomfort, how it would impact the routine activities, etc. There is also some understanding of various illnesses in the community one lives. For instance, a woman with bodyache and nausea, weakness is described and perceived to be quite normal by most women and would not feel worried.

In such social and household situation, Guddi's nausea and weakness were not considered important. She told, "My household members including my mother-in-law and Bhauji (husband's brother's wife) told me that most women would have these problems but it will not lead to any health breakdown. I should not feel low and stay active. Doing household jobs makes a woman physically tired but that does not mean she is unwell. But I was seriously not feeling well for more than a month. Later, when household members did not understand, I persuaded my husband to take me to Jharia, my natal place. I had already discussed it with my parents, and they asked me to come. There, I was taken to a hospital; I described my feelings and all past tendencies of nausea and stomach pain that were persistent, and the lady doctor asked me to do some investigation. As it was easier to do in a Bharat Coking Coal Limited (BCCL) hospital, followed as prescribed. The follow-up reports revealed that I had very low BP, 'Khoon kaame halau' (low haemoglobin), etc., that I do not understand how much, but 1 month of medicines were prescribed that my parents bought, and gradually, my health improved. I returned to Hadi Cooli one and a half months later when I was feeling better."

Guddi's illness was considered normal to women. Women do feel weakened, and it will be harmful to Guddi. However, she went to her mother's place. She was diagnosed with a tendency of low Blood Pressure (BP). The doctor provided medication and asked for follow up. Guddi felt satisfied about her treatment after consulting doctor in BCCL hospital at Dhanbad. She did not pay any amount; a minimum amount was charged to her father as he is employed at BCCL.

It was also because Guddi was a newly married and had a lesser understanding of the surrounding in Chas. She could not force her mother-in-law but now Guddi can go to a doctor all alone if such feelings of being unwell persist. She also has some amount that can be spent on visiting a doctor and buying medicines. Kosa and Robertson (1969) also discusses that the viewpoint about a disease is not independently seen and exists in a social situation but it also gets influenced by social factors. In his above-given illustration, the illness behaviour is not seen as an individual matter but deliberated and guided in the household and society where one lives. In brief, illness behaviour is also an outcome of social situations that helps to diagnose, if required, and then arrive at a say whether someone has a disease or not.

Light and Schuller (1986, p. 5) state, "Medical care and health services are acts of political philosophy. Thus social and political values influence the choices made, institutions formed and funding levels provided for health care." Chas town has just one hospital, a sub-divisional hospital, which caters to the large population. In another context, regarding disease and illness, the RMP in the Bauri Cooli does not have the expertise to differentiate between the two. The expertise of doctors is focussed on curing the patients. However, they claim to diagnose the disease when any Hadi patient visits the clinic and seeks treatment. It is based on symptoms narrated but not an anthropological distinction that is their limitation. In his book Kleinman (1988) differentiates between disease and illness. Every experience and suffering faced by the individual in the process of bringing their body to normal routine reflects the illness. This narrative shows extremely painful moments suffered by Jhagu during his TB sickness. Illness narratives proved to be an important method (Kumar and

Mahapatro 2010) also used in my research with the Hadi community while exploring the perception of illness and health among them.

At this juncture, I find it partially appropriate to study Frank (1998); where argued that people suffering from deep illness tend to tell three predominant stories, and he defines profound illness as that which is "Perceived as lasting, as virtually affecting all life choices and decisions, and as altering identity" (Frank 1998, p. 197). The three predominant stories include the Restitution narrative, the Chaos narrative, and the Quest narrative. The Restitution narrative exists because of the relative efficacy of medical professionals in diagnosing and treating physical illnesses adequately. This story follows a general linear trajectory, in which an individual becomes ill but, through medical intervention, regains health. This is a popular and pervasive account. Most people tend to assume that health will be restored, sometimes even in the light of a serious diagnosis, such as cancer or HIV. When medical interventions fail, the person might feel a sense of injustice or astonishment, leaving him or her angry, in dejection, or mistrust. The belief that a restoration to health will ultimately occur in the future and it is simply a matter of time before that takes place can follow these feelings. Some people cling to the hope of restoration until the end, as in Raju Hadi's illness of cirrhosis. In contrast, the Chaos narrative is filled with uncertainty, confusion, and a sense of powerlessness over events in one's life. It refers to the perception that everything is falling to pieces. For the listener, this can be the most difficult account to hear. Mishler (1986) in her volume acknowledges need of listening to the individual narratives in a perspective of qualitative interviewing. The general complaint of the Hadis in the above discussion is that the doctors at government hospitals did not bother to hear about their illness. The first interaction with doctors creates anxiety in the minds of Hadi patients.

3.9 Occupation and Illness Experiences

It is important to refer to the concepts of caste and occupation to understand the perception of illness and its experiences. Massey (1997) argues that the process of and internalising any event is an integral part of human behaviour, which influences life activity and chances. It is quite obvious that the myriad forms of discrimination due to caste and occupation (Khare 1984) shape the mindset of people like the Hadis and those residing in their surroundings as well. Hadis and other similar caste groups are forced to be attached to unclean occupations of cleaning latrines, gutters, etc., continuing as a caste-linked occupation. Despite going through several changes, the perception that Hadis are manual scavengers exists even today.

The decision to seek professional health care depends on past experiences and the socio-cultural context in the form of social relationships where an individual resides (Weiss and Lonnquist 1997, p. 135). The doctors in the locality of Hadi Cooli appreciate this socio-cultural context. The decision to seek help is important, but it is easier to approach the doctors of the locality than any physician at both private and public care centres.

DiMatteo and Freidman (1982) identify three factors which may influence the decision: first, the background of the patient; second, the patient's perceptions of illness; and last, the social situation. Zola (1973) also has identified five social triggers that influence the judgement of the patients to seek health professional needs: "First, perceived interference with physical activity (work related); second, perceived interference of social and personal relations; third, an interpersonal crisis; fourth, setting the deadline, i.e. if I am not better by Monday, I will call the physician" (Zola 1973, pp. 677–689). The fifth triggering factor is important among Hadi individuals. The pressure from the family and friends becomes important, but in addition, the importance of the sick person to the household is considered if he or she is the sole earner or if he or she is a small child. The households do not delay in the cases of children and seek treatment if seriousness is perceived. Dheeran Hadi shared his experience which is relevant in this context. When his younger child Rajni was unwell with a high fever, they thought that it was not a fever and ignored it the first night. True to his perception, she was alright by the next morning. But the next night, her temperature was again rising, and his daughter, who was just 6 years old, started mumbling. Rajni was very uncomfortable, and her legs and head were boiling. He did not take a chance, so he took a bicycle from his neighbour and rushed her to Muskan Hospital. They said that she was in a very critical condition, but thanks to the doctor who gave one injection and two tablets to her, and within 1 h, Rajni was normal. When she realised that she was not at home, she got anxious. Dheeran says my daughter started crying saying let's go home; she feared the doctors and patients in the emergency ward. After taking a blood sample, the doctor on duty asked Dheeran to take Rajni back home. The report would come next day and suggested she should come back. The next day doctor examined the investigation report and said that it was a minor ailment. It was pneumonia and viral fever. Two weeks medication followed and that relieved Dheeran because his daughter had recovered well. The entire experience was horrible but he did not delay or postpone the treatment. He informed that he had spent almost 2000 for her treatment as they had no choice. The parents rushed to Muskan because it provides emergency services to children. The doctors are available round the clock hence he preferred to go to Muskan hospital.

At times, people perceive and do relate illness experience with religion that God causes disease and illness. These mechanisms are used by people when patient in illness do not find any other alternative, they seek the cure in the name of God and attempt to heal themselves.

Overall, the illness process for Hadis begins at home but ends at several points, i.e. at the medical stores, at home, or the local doctor, PHC, or any other private clinics. I will explore these details in the next chapter. The following chapter also illustrates the fact that illness can be perceived through various means; some of them are discussed in next chapter.

References

Acharya, Sanghmitra S. 2010a. Caste and Patterns of Discrimination in Rural Public Health Care Services. In Sukhdeo Thorat and Katherine S. Newman (eds.) *Blocked by Caste-Economic Discrimination and Social Exclusion in Modern India,* pp.208-29. Delhi: Oxford University Press.

Acharya, Sanghmitra S. 2010b. Access to Health Care and Patterns of Discrimination: A Study of Dalit Children in Selected Villages of Gujarat and Rajasthan. *Children, Social Exclusion and Development, Working Paper Series,* 1 (2). New Delhi: Indian Institute of Dalit Studies (IIDS).

Acharya, Sanghmitra S., Mala Mukherjee., and Kanhaiya Kumar. 2015. Social Inclusion in Universal Access to Antenatal Care: Understanding the Inter Group Inequalities. *Journal of Social Inclusion,* 2 (4), New Delhi: Indian Institute of Dalit Studies (IIDS).

Apple, D. 1960. Sociological Studies of Health and Sickness: A Source Book for the Health Professions. McGraw- Hill. SAGE Journals.

Armstrong, D. 1984. The Patient View. *Social Science and Medicine.* 18 (9): 737-44.

Banerji, D. 1971. *Family Planning in India: A Critique and A Perspective.* New Delhi: People's Publishing House.

Banerji, D. 1974. Social and Cultural Foundations of Health Services Systems. *Economic and Political Weekly.* 9 (32/34): 1333-1346.

Banerji, D. 1978. Political Dimensions of Health and Health Services. *Economic and Political Weekly.* 13 (22): 924-928.

Banerji, D. 1981a. Measurement of Poverty and Undernutrition. *Economic and Political Weekly.* 16 (39): 1579-1582.

Banerji, D. 1981b. The place of Indigenous and Western Systems of Medicine in the Health Services of India. *Social Science and Medicine.* 15A (2): 109-14.

Banerji, D. 1981c. *Poverty, Class and Health Culture in India.* New Delhi: Prachi Prakashan.

Banerji, D. 1984. Primary Health Care: Selective or Comprehensive. *World Health Forum.* 5 (4): 312-5.

Banerji, D. 1989. Rural Social Transformation and Changes in Health Behaviour. *Economic and Political Weekly.* 24 (26): 1474-1479.

Baru, Rama. *et al.* 2010. Inequities in Access to Health Services in India: Caste, Class and Region. *Economic and Political Weekly.* 45 (38): 49-58.

Bhore Committee. 1946. A study of the health needs of the people and their implementation. Government of India.

Caudill, William. 1956. Perspectives on Administration in Psychiatric Hospitals. *Administrative Science Quarterly.* 1 (2): 155-170.

Census of India. 2001. Basic Data Sheet: District Bokaro- Jharkhand. (13), accessed 19 April, 2015. http://censusindia.gov.in/Dist_File/datasheet-2013.pdf.

Census of India, 2011. *District Census Handbook.* Jharkhand Series-21. Part XII-B. New Delhi: Registrar General of India.

DiMatteo, M. Robin and Freidman, Howard S. 1982. A Model Course in Social Psychology and Health. *Health Psychology.* 1 (2): 181-193.

Early, E. A. 1982. The Logic of Well Being: Therapeutic Narratives in Cairo, Egypt. *Social Science and Medicine.* 16 (16): 491-97.

Frank, Arthur W. 1998. Just listening: Narrative and Deep Illness. *Families, Systems and Health.* 16 (3): 197-212.

Gregory, Richard L. 1987. Perception. In R.L Gregory and O L Zangwill (eds.) *The Oxford Companion of the Mind*, pp. 598-601., Oxford: Oxford University Press.

Hyden, G. C. 1997. Civil society, social capital, and development: Dissection of a complex discourse. *Studies in Comparative International Development, 32*(1), 3–30. SpringerLink. https://doi.org/10.1007/BF02696304

Katavic, Snjezana *et al.* 2015. Illness Perception and Information behaviour of Patients with rare chronic diseases.

Keith, D. A., *et al.* 2021. Indicative distribution maps for ecological functional groups—Level 3 of IUCN global ecosystem typology, version 2.0.1b. Zenodo. https://doi.org/10.5281/zenodo.1234567

Khare, R. S. 1984. *The Untouchable Himself: Ideology and Identity*. Cambridge: Cambridge University Press.

Kleinman, Arthur. 1980. *Patients and Healers in the Context of Culture: An Exploration of the Borderland between Anthropology, Medicine, and Psychiatry*. California: University of California Press.

Kleinman, Arthur. 1985. Interpreting Illness Experience and Clinical Meanings: How I See Clinically Applied Anthropology. *Medical Anthropology Quarterly*. 16 (3): 69-71.

Kleinman, Arthur. 1988. *The Illness Narratives: Suffering, Healing, and the Human Condition*. New York: Basic Books.

Kosa, J. and L. Robertson. 1969. The Social Aspects of Health and Illness. In Kosa et al. (eds.) *Poverty and Health: A Sociological Analysis,* Cambridge: Harvard University Press.

Kumar, Ashish and Dilip Kumar Sinha. 2014. *Opposite Behaviour of Food Market from that of Other Commodities: A Cause of High Inflation*. A Working Paper, accessed on 22 Oct, 2016. http://mospi.nic.in/mospi_new/upload/paper.

Kumar, Avinash and Meerambika Mahapatro. 2010. Narrator and the Narrative: Understanding Scientific Realism of Chronic Diseases in the Field. In Arima Mishra *Health, Illness and Medicine* (ed.) New Delhi: Orient Longman.

Light, Donald and Alexander Schuller. 1986. The Political Values and Health Care: The German Experience. cited from William C Cockerham *Medical Sociology,* Englewood Cliffs: Prentice Hall.

Massey, James. 1997. *Down Trodden: The Struggle of India's Dalits for Identity, Solidarity and Liberation*. Geneva: WCC Publications.

Mathur, Indu and Sanjay Sharma. 1995. *Health Hazards, Gender, and Society*. Jaipur: Rawat Publications.

Mckeown, Thomas. 1979. *The Rise of Medicine: Dream, Mirage and Nemesis*? London: Nuffield Provincial Hospitals Trust.

Mechanic, David. 1986. The Concept of Illness Behaviour: Culture, Situation and Personal Predisposition. *Psychological Medicine*. 16 (1): 1-7.

Mishler, Elliot G. 1986. *Research Interviewing: Context and Narratives*. Cambridge: Harvard University Press.

Patro, P. K., and Acharya, S. K. 2016. Utilization of ferrochrome wastes such as ferrochrome ash and ferrochrome slag in concrete manufacturing. *Waste Management, 58*, 73–81. https://doi.org/10.1177/0734242X16654751

Petrie, K. J., L.A. Jago and D. A. Devcich. 2007. The Role of Illness Perceptions in Patients with Medical Conditions. Lippincott Williams and Wilkins: *Current Opinion in Psychiatry*. 20 (2-3): 163-167.

Qadeer, Imrana and Dunu, Roy. 1989. Work, Wealth and Health: Sociology of Workers' Health in India. *Social Scientist*. 17 (5/6): 45-92.

Qadeer, Imrana. 1995. Primary Health Care: A Paradise Lost. *IASSI Quarterly*. 14 (1&2): 1-20.

Qadeer, Imrana. 2002. Draft National Health Policy 2001-I: Debt Payment and Devaluing Elements of Public Health. *Economic and Political Weekly*. 37 (1): 12-16.

Reader, George. andMary E. W. Goss. 1959. The Sociology of Medicine. In Robert K. Merton., Leonard and Leonard S. Cotterll *Sociology Today: Problems and Prospects*. pp. 229-246. New York: Basic Books.

Rosenberg, Charles E. 1989. Disease in History: Frames and Framers. *Milbank Quarterly*. 67 (Supplement 1): 4-5.

Sahu, S. K. 1986. Social Dimension of Health of Tribals in India: A Case Study of Oraons of Orissa. In Buddhadeb Chaudhuri (ed.) *Tribal Health: Socio-cultural Dimensions*, New Delhi: Inter-India Publications.

Shah, A.M. and Aneeta Ahluwalia. 1970. Role of Sociology in Medical Education and Research. *Economic and Political Weekly.* 5 (17): 705-10.
Sigerist, Henry E. 1952. *History of Medicine.* New York: Oxford University Press.
Sujatha, V. 2003. *Health by the People: Sociology of Medical Lore.* New Delhi: Rawat Publications.
Weiss, G. L., and Lonnquist, L. E. 1997. The sociology of health, healing, and illness (3rd ed.). Prentice Hall.
Ziyauddin, K.M. 2022. Situating Hadis' Occupation and Caste. CASTE: A Global Journal on Social Exclusion 3 (2), 263-284
Zola, Irving K. 1973. Pathways to the Doctor - from Person to Patient. *Social Science and Medicine.* 7 (9): 677-689.
Zurbrigg, Sheila. 1984. *Rakku's Story: Structures of Ill Health and the Source of Change.* Bangalore: Centre for Social Action.

Chapter 4
Perceiving Health Care Services Through *Suya-Goli*

4.1 Introduction

This chapter explores what people in Hadi Cooli call *suya-goli*, suya meaning injection, and *goli* meaning tablets, and how most of the medication and care given in terms of health for Hadis revolves around these two things. The attempt here is to understand the Hadis' perceptions of the common practices of health care.

For Hadis, *suya* acts as an ultimate treatment for an ailment. The prescription of *goli* increases the level of satisfaction among the Hadis. Nearly everyone who visits the doctor's clinic in the locality finds a consolation prize in these elements, *suya* and *goli*. However, seeking health care services begins first at the *davaee dukaan* (medical shop) for treatment. Chas medical shop is quite close to Hadi Cooli and will be referred to and discussed in this chapter in the context of perceiving health care by Hadis. The next stage is at Doctor Vimal Singh's clinic in Bauri Cooli. The chapter also brings the perspective of Hadis employed at Bokaro Steel Plant (BSL) and at Chas Nagarpalika. This difference is noted only for the reason that the steel plant provides free health care services to employees, whereas Chas Nagarpalika has no provision for health care services for its employees. The employment of sweepers working at Nagarpalika is contractual. Therefore, they, like other Hadis, rely on low-cost treatment provided in the vicinity of Hadi Cooli.

Zola (1973) describes that the perception of symptoms is strongly influenced by the extent to which the symptoms interfere with the person's daily life, and lay assessment of the severity of symptoms is often judged against this criterion. This kind of interference in the daily activities of an individual sometimes creates trouble and pain. The symptoms that interfere with the person's everyday work are examined and understood through a lay assessment and how much it will interfere in their everyday life. The severity of the problem in the individual's body is not accurately perceived. Hadis, at times, do relate an illness to different factors that would be perceived wrongly when consulted with the doctor. In simple explanation,

K. M. Ziyauddin, *Occupational Caste Groups' Illness and Health in India*, People, Cultures and Societies: Exploring and Documenting Diversities,
https://doi.org/10.1007/978-981-96-8967-5_4

patients have their own way of explaining the discomfort that need not necessarily be medically authentic.

4.2 Postponement of Medication and Spending on Medicine

Hadis have so far developed a certain pattern of seeking treatment for their illness. The socio-cultural aloofness of the trained physician in hospitals, either private or public, constructs a wall-like barrier between a doctor and a patient except doctors in private or public, constructs a wall-like barrier between a doctor and a patient, except for doctors practising in the neighbourhood. This acts as a delaying factor in reporting illness at the Doctor Vimal Singh's clinic in Bauri Cooli, a neighbouring Mohalla. Here Hadis perceive the clinic as a private clinic of their doctor, an RMP. Hadis have been dependent on the treatment provided by Doctor Vimal Singh for more than a few decades. I have also mentioned the importance of a private clinic in the neighbourhood perceived as the referral centre among Hadis. For instance, Doctor Vimal Singh is also a resident living in Shakaro Cooli, a backward caste Cooli. He is conversant with the local culture and traditions and is aware of their health needs. Most castes, namely, Hadis, Bauri, Shakaro, Das, and Muslims, speak a common language—Khortha (one of the state's languages). There is a common connection between the castes through a common language and proximity of residences. In the overall familiarity, Hadis find it easy to approach medical shops and Doctor Vimal Singh in most illnesses easy to approach medical shops and Doctor Vimal Singh in most illnesses. Hadis interact with the owner of *Davaee Dukaan* (medical shop) even when he is not at the shop. He also lives in the Muslim Mohalla, Western side to Hadi Cooli.

The feeling of discomfort and pain or body aches and inability to carry routine jobs creates hindrances to attending work and is seen as an illness. These internal feelings enforce a Hadi to think upon the symptoms experienced and give certain meaning in order to seek treatment. Suchman (1965, p. 114) stages of illness and medical care determine five stages that differentiate critical transition and decision-making points in medical care and behaviour. In the first stage, a patient feels that there is something wrong. This stage provides three patterns of experiences, firstly, having pain, discomfort, feelings of weakness, and feeling of inability to do routine activities. Secondly, when a patient derives certain meaning out of the discomfort and pain in the body and thirdly, a kind of emotional response to the anxiety due to pain and discomfort faced. In all, the condition of the body is defined and understood by an individual, and certain meaning is provided to express the level of illness and seek treatment. The emotional response of anxiety that comprises the patterns of pain or discomfort and giving meaning to experiences is observed among the Hadis also. However, the Hadis may not always seek treatment.

In the second stage, an individual accepts that pain and discomfort are the symptoms of an illness that need proper medical attention and diagnosis. This perception helps the individuals to give up their routine work and switch over to adopt a sick role. A change in the role of individuals takes place from normal to a sick role in order to receive the medical support system. Therefore, in this stage, the seriousness of the perceived symptoms brings a transition from a normal to a sick role.

In his third stage of medical care, contact/self-care, Suchman (1965, pp. 114–116) described it as a point when an individual seeks medical care. However, medical sociologists put forth the notion of medical care having a variety of options at present. In this situation, patients seek medical care attention for the illness perceived and diagnosed. DiMatteo and Friedman's specification that the following three factors influence the decision to seek care are; (1) the background of the patient (age, gender, race, ethnicity, and class); (2) the patient's perception of an illness that would trigger the decision according to the symptom and seek professional health care and (3) the social situation of the patient. What is important to the patient on the day, whether he should attend work and postpone the reporting for other days, is decided accordingly. The priority on those days is important, and if possible, patients may postpone the decision to seek care.

The fourth stage is known as the dependent-patient role, where a doctor is required to make all the possible best efforts to help the patient return to normal routine and role. Lastly fourth stage is known as recovery and rehabilitation. There is the possibility that patients would return to normal life after getting proper medical support and medical care. There would be a few patients who would enter into the fifth stage of recovery and rehabilitation. Patients are provided specialised treatment for chronic illness, and later, they need rehabilitation to return to their normal activities.

Suchman (1965) states that the perceived seriousness of illness helps people determine whether to seek treatment, or they develop mechanisms to delay seeking treatment. In these situations, Hadis either seek treatment out of pain and discomfort or deny the feelings of pain to delay treatment seeking. Hadis is found to be delaying the reporting of illness at the first stage and later buying medicines over the counter.

4.3 *Suya*: A Quick Healing Method

The conceptual explanation to understand health care perceptions through *Suya* is pertinent with having *goli*. Hadis buy *goli* at Chas Medical Shop. My attempt to elaborate on the health care perceptions among Hadis appeared at three levels on health care perceptions among Hadis appeared at three levels, as discussed in this chapter described later. In the first stage, Hadi's effort is to overcome the perceived illness in the home setting. Households do manage and seek home remedies in order to minimise health expenditures and loss of wages if they want a consultation with a doctor. The overall perception at the personal level does not intend to seek the help of *goli* and *Suya* until home remedy is not helping. The perceived inability of home remedies to control any illness and when personal efforts are sensed to be failed, the illness level pushes Hadis for remedies to control any illness, and when personal efforts are sensed to be failed, the illness level pushes Hadis for the second level of treatment. In its second stage, their health behaviour of visiting a *davaee dukaan* to buy *goli* and manage the everyday routine of life is common. The third level a local doctor in Bauri Cooli a local doctor in Bauri Cooli's to visit a local doctor in Bauri Cooli, which informs the stages of medication commonly adopted by Hadis in dealing with their illness. Some occasions influence and guide Hadis to seek treatment from the doctor. Dr. Vimal Singh is their trustworthy doctor who is approached for consultation and medication. A good number of them had already taken medicine over the counter. They usually tell the doctor about the medicines they have already taken. When one set of providers appears not to be helping to cure or minimise the illnesses at *davaee dukaan, they* shuffle from *davaee dukaan* to a clinic. This shift is another stage in seeking better treatment. Hadis are aware that Suhail, at *davaee dukaan*, is not a doctor, but he can provide *goli* to cure, and it often helps to manage a problem (sometimes temporarily). I have discussed Suhail's narration on his know-how about medicines. It is sometimes as good as that of a RMP, he claims. Hadis shift their centre of treatment from *davaee dukaan* to a doctor's clinic when *goli* is perceived as not helping. An additional component is when the shop owner also sends patients to a doctor's clinic, but he does not enforce whom they should consult. The studies on injections give in-depth descriptions of how people perceive it in the process of seeking health care services. The role of *suya* (injection) in providing care to patients is one of the significant perceptions embedded in the socio-cultural milieu of Hadis' life. The popular perception is that suya would cure even prolong illness, and one does not need to disbelieve this notion. The process of seeking treatment through *suya* is enormous and socially significant. They attach extra importance to *suya*. *Suya* is more costly than *goli*. I have studied *suya* as a means of pain relief and cure for many ailment. Suya is believed as productive by the Hadis. I have used *suya* as an ethnography of Hadis in Chas, provided mostly at Dr. Vimal Singh's clinic. Pari Devi, aged 42, is a sweeper in a hotel near Chas bus stand market. She is a widow. Her son Bablu Hadi and daughter-in-law Sonam live with her along with her grandson Veera. Her husband, Raju Hadi, passed away due to chronic liver cirrhosis. Pari perceives *goli and suya* as helpful ways to and minimise the many an ill feeling. She said, "In severe backache and weakness, even one *suya* helps me return to normal work as the pain reduces and even goes away. Doctor Vimal Singh generally asks whether he should administer a *suya* because it involves higher charges than having a *goli*. In a condition of 'unbearable discomfort' (*khoncha jae hau deha)* in an illness, I would request him to give *suya*. However, the final call is taken by a doctor as he is a doctor."

In the paper written on 'Injection practices in the formal and informal health care sectors in rural north India', Kermode and Muani (2006) have described that injections are not only provided by trained or authorised practitioners such as doctors and nurses but also by large number of informal practitioners such traditional healers. The study focussed on patients from one rural hospital, each in Bihar and Jharkhand. It analysed the trust and over-dependency on injections (*suya*) from informal practitioners. This study followed a cross-sectional survey comprising 280 patients and 120 relatives of the patients, contrasting my ethnography on a small group in Chas. As their sample was in-patients, 43% of them visited informal practitioners and had received at least one injection prior to their admission to the hospital. The higher proportion of the rural population in Bihar and Jharkhand taking injections shows the dependency on local informal practitioners who provide injections. Forty-three percent of their study participants seem to have thought it was their best and most effective way of treatment. Even among relatives of the patients, 13% of them received at least one injection in the last 3 months. The role of informal practitioners is quite visible in terms of seeking health care in the study by Kermode and Muani (2006). Similar patterns towards health-seeking behaviour and perception in favour of injections are also common among Hadis in Chas town who seek care from injections once *goli* does not help.

Another study describes the practice of injections by untrained practitioners due to the socio-cultural and financial conditions of the patients in India (Reeler 2000). Anand et al. (2001) noticed that injections act as the most effective way to give satisfaction to patients as it is perceived and overvalued in comparison to oral medication. Health practitioners prescribe the injections as it is the demand and perception of the patients that has built over time in India. In another study, it was observed that a higher percentage of the sample (46) received an injection, at least in the selected urban population based in New Delhi (Kotwal et al. 2004).

Mahapatro and Kalla (2000), in their paper on 'Health Seeking Behaviour in a tribal Setting', described that villagers visited doctors' residences during the night and paid Rs. 25 rupees per injection in emergency conditions. They found the perception of the people that if injections are given, they will be cured, and it is treated as the best way for quick relief. This finding emerges from the study of six tribal villages where 621 women were studied. This study brought 'quick therapy' as the mechanism among the villagers, which comes from injection and not oral medication. Almost similar to Hadi households, they also delay illness reporting at the first instance and utilise a home remedy. When the illness is not treated by home remedies, they visit practitioners. Hadis do not stand similarly in terms of traditional healing. The studied tribal villages (Mahapatro and Kalla 2000) have extensive availability and use of traditional medicines which does not appear among Hadis. These villagers also believed the presence of the goddess in the human body was similar to the goddess Sitala among Hadis. Measles (*gundi*) and chicken pox (maa) are not only perceived to show the presence of the goddess but they are also believed to be the wrath of the goddess (Thakur). This health belief takes them to '*desari*' (healers). In very rare cases, Hadis sometimes seek help from a Baba and *Ojha* when illness is not treated through allopathic facilities.

There is a direct implication of *suya* to cure the illness of Hadis. Several Hadis narrate it, "*Suya liye ke baade ab baes bhae jeebae. Katae sahbin ee beemariya hau.*" Literally, after taking an injection, I will be alright. How long could you bear the illness? It is in general discussion that suya is perceived as a strong and effective means to control and cure prolonged illness. Hadis counted the number of suya injected to show the gravity of the illness suffered. I recall in one of the interviews, Simar Devi, a widow, said, "My husband (Jhagu Hadi) was given around 10 *suya* to cure his cough (*chatia*). We were always hopeful that *Suya would* help him to recover properly. Each time the doctor gave one *suya*, he could feel some relief in his chest pain (*chatia*), which was due to constant cough. Perhaps Mansa Devi had written that much of life to him." Simar, age 65, is the widow of Jhagu Hadi. She resides with her son Anulup Hadi and reported that they tried seeking better treatment from Dr. Vimal Singh. He was not only given *goli* for almost 2 weeks, but later on, *suya* was given for 10 days administered. In most illnesses, she recalls *suya* cured her problem. She said having cured herself by suya whenever she had prolonged weakness and body ache. Dr. Vimal Singh usually charges 30 rupees for a *suya* whenever she asks. Simar has also seen her friends finding relief from illness when injections are given. How it did not provide relief and cure to her husband is beyond her understanding. She feels her husband lived a life blessed by goddess Mansa. The doctor also tells her that he had changed the medicine in *suya* to improve its effectiveness, but even then, it didn't cure her husband.

It is also relevant when Hadi feels indecisive about the health care providers. The process of seeking health care gets sanctity when Doctor Vimal Singh prescribes further treatment. In cases of some medication, Doctor Vimal Singh advises for follow-up treatment, and he tries to explain the benefits of the *suya* to the patients. In an interview, Doctor Vimal Singh said, "Hadis are poor. I advise them to visit Chas Hospital as it is a government hospital. They cannot afford expenses on medical treatment at K M Memorial Hospital, located nearby, at Chas Bus Stand. However, I have seen that they generally do not go to KMMH or Chas Hospital despite my reference to visit Chas Hospital. When I inquired if they went to the hospital (KM Memorial Hospital-KMMH), they gave some excuse and sometimes plainly mentioned their inability to pay the high fees that delayed the hospital visit." Hadis do not prefer to visit big private hospitals like K M Memorial and Muskan Hospital. Though KMMH at Chas Bus Stand is at a distance of just 1 km, Hadis preferred and chose to visit the RMP's clinic.

In the above situation, it appears that Hadis got suggestions from their doctor, Dr. Vimal Singh and also discussed visiting a bigger hospital at home. They do not immediately visit an expensive hospital even after Doctor Vimal Singh advises them to do so. The cost of treatment is relatively high in big hospitals. The consultation fee to a physician at KMMH is rupees Rs. 250. Hadis feel they pay less than 100 rupees to Doctor Vimal Singh and find effective treatment. For this reason, sometimes they ignore and delay further treatment if advised to visit the bigger and more costly hospital. It consequently increases the morbidity of Hadis. Doctor Vimal Singh informed me that when patients have a high degree of discomfort and feeling of being unwell, I provide *suya*. It provides immediate relief to them. Therefore, *suya is urgently required for those patients who have prolonged illness, and it is decided by the* doctor. I Doctors like Vimal Singh and Ganesh know rather well that Hadis needs to get back to work and cannot afford to stay away from work. They know that an injection would work faster. Hadis themselves perceive the speedy recovery through *suya*. The RMPs (Dr. Vimal Singh and Ganesh) know that the Hadis try treating the problems with *goli* bought over the counter before approaching the doctors. Thus, the doctors believe they need to administer *suya* for quicker relief. *Suya* plays a pivotal role in the health care-seeking process of Hadis, especially after they have tried a cure through home remedies and *goli*.

Lamti Devi, aged 64, feels weakness and occasional body aches. Normally she does her routine household chores. She helps her daughter-in-law in cooking. She would cut vegetables for curry and oversee the preparation of rice that does not demand a heavy workload. Lamti does not like winters. It increases her knee pain. She manages with *goli* sometimes. But at times, the pain is severe, and she cannot bear it. Last year, in 2011, Lamti told me I took *suya* for my weakness and cough in the winter. I was given five *suyas* in 2011. While describing her dependency on *suya*, she emphasised on the number of *suya* given to her. Once, she said, "My son gets a few painkiller *goli* from BGH. Doctor Vimal Singh would explain me the doses. I have been to BGH, it is very crowded. The person at the *davaee dukaan* at BGH does not get time to talk any patients. Therefore, I avoid visiting BGH. Rather, my son gets painkillers and goli for weakness from the hospital on his card whenever he visits for his own treatment." Bigger hospitals like BGH prescribe oral medication to most outpatients until it is a critical case. In contrast, Hadis have a higher dependency on *suya* than oral medication. Injection for weakness and body pain is given at the cost of 50 rupees per dose. Ramesh once said, "I pay 50 rupees to Dr. Ganesh for a *suya* given at my home. It helps me to stay active and minimises pain due to heart ailments.

The trust in *suya* is almost natural. It works quickly and provides relief in illness and pain. Doctor Vimal Singh also provides *suya* after childbirth for tetanus. He said it is safe to give *suya* after childbirth. He provides measles, mumps, and rubella (MMR) *suya* to some women at the last stage of pregnancy. But it cost 400 rupees and more that is perceived quite high amount by Hadis. Medical representatives request Dr. Vimal Singh to use MMR injections. To Hadis, *suya* is considered crucial when in illness. Pregnancy and childbirth related suya is not perceived as essential.

Lamti Devi said, "*Suya liye mein eke dikkat hau, ek bari chaalis taka le lae hau kintu kankania to choo mantar bhae ja hau.*" Suya acts as a quick healer, but it costs 40 rupees for a single dose. It heals quickly, like good magic. In medical emergencies, pain reliever suya works for a few women. He gave a few names of medicines used in *suya*, pethidine or dimorphine; opiods are a few names of medicines he injects in the thigh or buttock. Most women deliver without injections. However, several houses call him at the time of childbirth.

Chalni Devi, aged 50, takes *suya* whenever she feels weak for a few days. During a visit, I observed that she was given *suya* in 2013 and administered one saline. She said, "I fall down if food is not taken on time. I feel weak immediately and cannot control myself. The doctor tells me it is an illness called BP. It will always remain a problem for me. Whenever I have feelings of being unwell and unable to do my everyday work, my son Ganesh will ask me to consult Doctor Vimal Singh. My son pays the fees and cost of *suya-goli*. Today is the second *suya* given to me, and he asked me to revisit it after a day. I do not know his charges; better you also ask a doctor."

The trust in *suya* is socially constructed. Ashish Hadi, aged 61, got medicine for his low blood pressure from his employer's hospital, BGH. After retirement, he does not go to hospital on a regular basis. Whenever he feels unwell for a few days and calculates the cost of treatment increasing at Doctor Vimal Singh's clinic, Ashish plans to visit BGH. He, along with his wife, Sani Devi, goes together sometimes. Once, Doctor Vimal Singh said, "BGH employees consult the doctor and tell many complaints to the doctor. Ashish gets tablets and a few doses of injections for his wife. She has a problem in her back. Constant pains trouble her. I give *suya* to her when she cannot do household chores. I use her medicine given at BGH and charge only 20 rupees for my consultation to visit her at home." Ten years earlier, Doctor Vimal Singh would use a syringe for several patients. He would ask the household member to boil one glass of water. Then, he would put the syringe in the boiled water in a pot. After 5 min, when he felt it was washed and cleaned, he would again reuse the same syringe. At present, it is quite easy and convenient for him to use disposable packets for injection.

4.4 Davaee Dukaan

Women prefer sending their children to the shop for medicines. Body pain, vomiting feeling, headache, cough, and weakness are quite commonly reported illnesses at the shop. I had several occasions to observe and even chat, although briefly, with those children. Women also seek medicines to subside or minimise their pain and discomfort. In a situation when an illness disturbs everyday and prolongs for a few days, Hadis would decide to consult Doctor Vimal Singh. Visiting his clinic does not require extra effort from women, unlike going to government hospitals, which needs longer commuting, transport costs, and half a day. One can walk oneself to his clinic and seek consultation. Malt Devi said, "We need not depend on our menfolk to meet Doctor Vimal Singh. It is convenient and comfortable to explain health problems. I pay even half of the fees. Sometimes, he reduces the fee. I remember when he gave me one bottle of syrup. I am sure he is a good doctor and a person as well. His treatment helps at most times."

The reporting pattern and forms of illness vary among them. Due to the health care benefits of having employment in Bokaro Steel Plant Limited (BSL) and Mines Area Development Authority (MADA), the BSL employee's households may not delay seeking medication in comparison with other Hadis. Bokaro General Hospital (BGH) is one of the best hospitals in the Bokaro district under BSL. Employees who have retired and the two other Hadis presently employed at BSL feel privileged to have a good hospital facility. Somesh, Ramesh, Shailender, Ashish, Pintu, and Manik reported that seeking treatment at BGH is bonus to the poor, like Hadis, that is not available to the rest employed in other sectors. Manik said, "I do not visit BGH for minor illnesses, located far away at 12 kilometres. It may incur higher expenditure than buying a few tablets at the *davaee dukaan*."

Chas Medical Shop attracts a good number of Hadis seeking tablets to cure or minimise minor illnesses. The observation at the medical shop provided more children seeking *goli* for the women at home. It is common and convenient to send young children to get some medicine on the basis of providing an account of illness felt. The habit of buying medicines at the medical shop helps to control the illness felt and also minimise the expenditure on treatment. Even sitting often for long hours at the clinics of the local doctor provided considerable data on Hadi's perceiving illness through *suya* and *goli.* Many Hadis are habituated to continue in some pain and discomfort. It is not considered a threat until work and household activities begin to be affected. Most of them try not to delay at work. Panbabu said, "*Hamni ke hi kaama emergency wala hau, nae karbin to officva aare roadva mien jam laeg ja hae. Afsavan bihaane jam dakhtun to pucha-paachee kare lago hathin hau.*" The works of Hadis are most urgent at all offices. Delays in collecting garbage from roads will block the traffic. The officers will realise the traffic when they go to office and they may inquire about it.

Hadis employed as sweepers in BSL had higher incomes than the rest of the Hadis, which also affected the decision to seek health care. Steel plant employees, on average, earn not less than rupees 12,000 per month. An average of Rs. 3800 rupees is paid to Chas Nagarpalika employees. The value in Hadis' perception is higher and worthwhile due to the satisfaction received at Davee Dukaan and in the Bauri Cooli clinic. No deterrent stops them from visiting Davaee Dukaan and Dr. Vimal Singh's clinic for health care needs. Even a child easily buys tablets on behalf of a woman at home without a prescription. It shows that Hadi women have confidence and belief in the efficacy of such given tablets at *davaee dukaan.* Many Hadis have been taking tablets for years and they find it helpful to minimise the discomfort. Bineeta Devi, 50, works in a house at Ram Nagar Colony. She walks for 20 minutes to reach her employer's house. Walking almost 40 minutes every day makes her more tired than doing the cleaning work. She said, "Every day I walk a long distance. Whether it is extreme summer heat or monsoon rain, my employer gets angry when I take leave for a day. Even when I am not well, I think of going to work. When there is extreme pain in my body, especially my feet and legs, I send my grandsons Prem, aged 8 and Vinod, aged 6, to get some tablets at Chas Medical Shop. In conditions of prolonged pain, she reported to Doctor Vimal Singh in his clinic. Another fact that needs to be mentioned is that the medical store is culturally familiar to Hadis and distance-wise closer. The Hadis who take a few tablets in order to maintain their normal activities and cure illnesses are guided by a medical shop owner and provided certain tablets. The trust between *davaee* and Hadis is culturally significant. The medical shop owner does give patient hearing to Hadis, after carefully listening what is told to him. Suhail, the medical shop owner, informed the concerned Hadi about the medicine, the cost of tablets, and the doses to be taken while dispensing the medicines. This description shows the social interaction between the two.

At the store, Hadis does not act as a patient suffering from a certain degree of illness. At the medical shop, the druggist and the Hadi patient do have social interaction similar but not the same as between a doctor and a patient in a clinical set-up. Suhail (*davaee dukaan* wala) does not act as a doctor, rather he engages with Hadis as Suhail, whom most Hadis know as a neighbour and a medical shop owner. This difference allows Hadis to narrate their minor illness and directly seek some medicines. Having such a socio-cultural mutual understanding enables Suhail to ask any Hadi visiting his shop, "*Ki bhalao, Ganesh.*" *Literally*, what happened to you, Ganesh? And he would dispense *goli* according to the money in the Hadi's hand.

In another encounter, one boy sought tablets for his mother. The person buying medicine on behalf of a sick person mostly finds a way to explain, in detail, about the symptoms of illness. Suhail asked a few questions of the boy about his mother's illness. Is her body warm? Does she feel cold and shiver? What about cough? The boy and Suhail's encounters show familiarity with each other. He knows the most Hadis by face. A child has no inhibition to explain his mother's illness to Suhail, which may not be the case at a hospital.

4.5 RMP in the Vicinity

There is a feeling of having an interpersonal relationship between the doctor and Hadis. A Hadi patient finds comfort in approaching a doctor who recognises her/him. Doctor Vimal Singh and Doctor Ganesh practice mostly with patients from lower caste houses. The clinic of Doctor Vimal Singh is in the midst of amid lower caste Bauri Cooli, as shown in Map 1 of Hadi Cooli in Chap. 2. Bauri Cooli is on the Northern side of Hadi Cooli. Shakaro Cooli is located on the Western side, and the Southeastern side has Muslim houses. Most households in these Cooli and recognise the elder Hadis, and Hadis too know them. It is told by a few Hadis that Doctor Vimal Singh has known us for years. He is not a person who would maintain a distance and treat us unequally. He talks in the language that Hadis also speak. There is no formality or procedure to visit his clinic, unlike any hospital. The doctor behaves nicely and provides a sense of comfort that allows Hadis to narrate their illness-related problems. The payment of fee is not demanded in advance. One Doctor Vimal Singh said, "My patients are aware that I am not going to charge an extra fee. If they have cash in hand, I get my fee on time. Otherwise, it is paid as per their convenience. Whenever I need some money urgently, I do, and I ask those whose dues are pending. Patients respect me, and in return, I respect them. They are not only patients, but I know them personally as part of a common neighbourhood."

The recognition of a patient by the doctor provides space of mutual respect that is mutually acknowledged. It also increases better communication between the doctor and the patients and gives a nuanced and personally embedded relationship between Hadis and the doctor. The cultural proximity between the doctor and Hadis helps to create a space where there can be a free flow of communication. In other words, the common linguistic association in *Khortha* binds them in one thread and strengthens the trust of doctors. Similar trust is minimal at public and private hospitals except for practitioners like RMPs, who have a close association with practitioners like RMPs, who have close association with Hadis and other castes. The present analysis is to RMPs as Hadi's primary point of contact in the locality.

In fact, with the local doctors, Hadis to describe the illness quite confidently, unlike the interactions with a physician (qualified) outside the locality in any of the hospitals, where there are other barriers. Non-familiarity with a hospital set-up, where meeting a doctor demands prior appointments at some doctors and hospitals, advance payment of consultation is a common practice. Sanjini Devi, 50, said, "I cannot walk such a long distance to visit a *SarPari Dactor* (government doctor). I already get regular backache. There is no certainty to meet a doctor even if I go. My husband tells me several stories about when our Hadi fellows went and could not consult the doctor. It is quite easy to find Doctor Vimal Singh every day in his clinic. In an emergency, we can request him to make a home visit." She has known Doctor Vimal Singh for several years. He understands the financial burden of Hadis. The fees are minimised if requested. The availability of government hospital is a little far away from Hadi Cooli at a distance of six kilometres. No sub-centre functions location is closer to Hadi Cooli except PHC at a distance of three kilometres.

Two points of view appear first: doctors (RMPs) and druggists usually know the neighbourhood households. There is a trust that these people will not run away after paying partial money or fees. Doctor Vimal Singh said, "In case there is a delay in making payment, I will ask the patients to pay whenever I am in need." Hence, the trust between Hadis and doctors and druggists sustains and carries forward. Second, Hadis are also aware that in situations of urgent health care, treatment can be availed, and the payment can be made later. This delayed payment does not humiliate Hadis while seeking treatment, unlike at private hospitals. This kind of trust and untold understanding makes the relationship smoother and more sustainable.

There have been situations when Hadis could not pay fees for quite a few months. For instance, Mooni Devi, 45 a 45-year-old widow, recalls how her husband worked hard to earn his livelihood despite being unwell for long periods. Her husband, Suraj Hadi, died due to prolonged illness of typhoid fever and other undiagnosed diseases (perhaps TB). She lives with her son and daughter-in-law. All three grandchildren love their grandmother and keep her busy during evening hours. To support the household, Mooni Devi also found a job at Punjabi Hotel in the nearby market area. She earns Rs. 1200 per month. They saved Rs. 15,000 rupees when her husband was alive. But the saving was used for his treatment in his last years of life. Her son Sanjit Hadi is employed at Nagarpalika and earns rupees Rs. 5700 monthly. The possibility of households being pushed to Below the Poverty Level (BPL) every year in India is quite due to out-of-pocket expenses like Mooni Devi. According to a study conducted in 2010, the number of people falling BLP is higher than 63 million due to increased spending on health, almost 7% of the country's population (Shahrawat and Rao 2011). The report of the Ministry of Health and Family Welfare, GOI shows that 18% of households faced disastrous health burdens in 2011–2012 as compared to 2004–2005 (Burman et al., 2010). The most health impoverishment is seen in rural areas when they seek outpatient care (OPD) (GoI 2014; Doorslaer 2007). Garg and Karan, in their paper on 'Reducing out-of-pocket expenditures to reduce poverty', analyse that the largest proportion of the out-of-pocket (OOP) money is spent purchasing drugs and the least on any in-patient treatment by poor households. They describe that the purchasing proportion of drugs is higher and shows higher drug sales in poorer states. The study finds OOP expenditures of 74% in rural areas and 67% in urban areas are drug-related. In the poorest states, the proportion of drug sales is even higher (Garg and Karan 2009).

The local doctor's health services are of utmost importance to the poor Hadis. Sonam Devi, wife of Kabir Hadi, reported, "Doctors like Vimal Singh are doing justice to his profession. He provides consultation to us even without charging fee sometimes. It depends on our relationship with him. The doctor has provided life-saving centre for us. He makes patients feel at ease and comfortable during consultation. He would inquire from people even when he met them on the way outside the clinic. That is never possible with doctors in Chas Hospital and Muskan Hospital. He and many others in the surrounding locality serve the people at their doorstep with the bare minimum fees (charges) for treatment."

Besides, at Dr. Vimal Singh's clinic, Hadis visit and may also talk about their personal issues. Some patients would narrate the illness much later and keep discussing the matter of family, friends work, etc. I observed many men consulted stomach-related ailments. The doctor finds this problem due to the food habits of Hadis. He said, "Most Hadis do not eat breakfast and go to attend their work. Sometimes, they would eat outside. When they return home empty stomach, they eat something at home at lunchtime. This increases the gas formation in the stomach and causes problems related to intestinal infection. Most of them drink alcohol that is locally made and sold in Bauri Cooli. Hadi men also eat very spicy *Pakodi* (snack) while drinking. It all adds to their stomach ailments." They do not consult him for one kind of illness. Rather, patients consult him for most illnesses at his clinic. During my time spent in his clinic, fever and body ache were quite commonly reported illnesses across ages and gender. Fever is a most recurrent problem for Hadis. Bokaro is one of the high malarial zone districts in the state, as told by the district malaria control officer in 2010.

Hadis' consultation had shown a large number of fever cases in the clinic of Doctor Vimal Singh. Several rashes and other skin-related problems were found. The doctor narrated, "Mostly cuts and bruises are due to the work of Hadis. They do not use ointment and clean the cuts. Garbage would have injurious waste like blades and broken glasses. Hadis perceive such problems as a minor threat, but you can see it troubles them a lot at work if they get some wounds in their feet. But it is their habit. What can I do except provide medication." Their illness reporting pattern at the clinic does provide a picture of their common health problems. A Hadi man faces additional injury at work, collecting garbage from the roadside, cleaning drains, and entering maintenance holes in the town area whenever sewage drains are clogged or choked. In a situation those two Hadis employed at BSL have a lesser risk of injury than the Nagarpalika employees. The BSL provides well-crafted shoes and hand gloves to its sweepers. The safety measures help them safeguard from cuts, but the same safety is unavailable to others. Somesh, a BSL employee, said, "Nagarpalika can never match the facilities provided by Steel Plant. We had a supervisor who would monitor for safety at the workplace. In case of some cuts and other injuries, employees are immediately sent to BGH. However, who cares in Nagarpalika. Hadis does not have the relaxation to take off from work for a day in case of illness or injury. So Hadis consider such problems as quite normal health challenges. Whenever needed, they seek medicines from Doctor Vimal Singh and continue to work."

Hadis do perceive that private hospitals in town also provide treatment in situations but relatively at higher expenses than Doctor Vimal Singh. The cost of treatment also matters a lot to them, which guides towards Doctor Vimal Singh's clinic. Despite paying substantial amounts to doctors in private hospitals, Hadis does not receive satisfactory treatment, unlike Dr. Vimal Singh, their regular doctor in the neighbourhood. Sonam Devi, aged 22, resides with her mother-in-law, Pari Devi. She is married to Anulup Hadi. Anulup, aged 25, works at Nagarpalika. They have one son, a year and a half old, Veera. Sonam brings social dimension in seeking treatment at clinics. She pointed towards the physicians' behaviour that some respect and courtesy should be shown towards patients as well. A doctor should behave equally with most patients, as is the case at the clinic of Doctor Vimal Singh. Anulup Hadi pointed to the behaviour during a doctor's meeting with patients also varying for other reasons more than caste. He said, "I also help Doctor Vimal Singh sometimes. If there is a major problem at his house, whether choked pipes, blocked drain, etc., the doctor would call me to clean it." Doctor Vimal Singh has not employed any cleaners in his clinic, and he cleans himself. In certain situations, even Hadis help people in getting the menial work done at his house.

Somesh Hadi, a retired BSL employee, compares the facilities of BGH, where he was posted, and the primary health centre or Chas Sub Divisional Hospital located in the town. He said, "It is difficult to get treatment at the government hospital in Chas compared to BGH. BGH provides free treatment to or employees, both in-service and retired employees. But travelling to the BGH costs not less than 50 rupees in a shared auto. Hiring a taxi is not in Hadi's consideration. Not only the cost, but Hadi patients have to spend the entire day availing the hospital care services even if it is free. Similarly, the cost of BSL staff availing and accessing treatment at BGH is relatively high for retired Hadi employees who need to visit BGH for free treatment."

Lamti Devi, aged 64, and wife of Ramesh Hadi, said, "I do not prefer visiting BGH. I sent my son Manik to bring medicines for my husband, who is on regular medication due to his ailments (heart blockages and weakness). Until it is too serious a condition, we visit Doctor Vimal Singh." The doctor pays more respect to the Hadi patients employed at BGH. He knows that Steel Plant provides treatment to employees. They visit Doctor Vimal Singh for two reasons: First, to save on commutation costs; second, it will take almost half a day to seek consultation at BGH, even for a minor illness. Whereas it is easier for Hadis to seek consultation at Vimal Singh's clinic, which helps to save Hadis from both the above hassles.

It is also true that Chas town, including Bokaro, has good health care facilities. BGH is one of the best hospitals to provide treatment for most diseases. Many patients from neighbouring districts like Dhanbad, Giridih, Ramgarh, etc., are referred to BGH. The presence of many trained physicians makes Bokaro district better than many of the districts in the state of Jharkhand. However, most of the practising doctors have their clinics at Cooperative Colony within the larger area of the steel sector near Sector 12. Many of the doctors who are employed at BGH also have private practice. Secondly, the trend to access private clinics is focussed on Dr. Vimal Singh's clinic due to the non-availability of a district hospital until my stay in the field. However, there was only one Hadi patient, Kajal, aged 16, who visited a doctor at Cooperative Colony (as discussed in Chap. 3). This place is also quite far from Hadi Cooli, at a distance of 6 km.

Visiting a doctor in neighbouring Bauri Cooli seems quite accessible for Hadis than any government (public) and private health care centres in the town. People feel quite convenient as it is a five-minute from Hadi Cooli and less expensive. Dr. Vimal Singh helps Hadis to manage their activities and attend work. However, it is important to understand that the available health care providers are beyond the ambit of public health institutions.

It is noteworthy that the clinics do not serve as primary health care providers but rather as secondary providers, due to the series of steps involved in seeking their care before Hadi visits them. Mostly, bodily discomforts and problems are discussed at the household level. Some patients seek medicine over the counter at medical shops to subside their illnesses. In the process, Hadis report their illness to Doctor Vimal Singh in his clinic if the illness is not cured. The utilisation of health care services constructs the health and illness perceptions of Hadis while seeking treatment either at a medical shop or the clinic. In the first step, it is observed that Hadis perceived that it was better to take tablets from Suhail at his medical store. They feel taking two or three tablets will set them right. What one of them has to say rightly supports this fact, "*Doo teen goli khaibin to thik bhai jeebin" literally* take two-three tablets, you will be alright. This statement is common among the Hadis. Tablets are mostly available over the counter at the Chas Medical store in the vicinity of Hadi Cooli. The shop is nearer to the clinic of Doctor Vimal Singh from Hadi Cooli.

There are three medical shops located at a distance of five-minute walk from Hadi Cooli. A Hadi visits a medical store for an illness. The illnesses reported to the medical shop owners are minor, but these have a major effect on their work, daily routine, and employment activities. Mantu Hadi narrated, "My illness, such as bodyache and cold, is minor, and I can tolerate and cope with it. But if this minor illness is not reported and no medicine or tablet is taken, then I may have a health breakdown (*Toote Ja hiye*). If I am allowed, I will take rest and overcome the illness, but who will manage my work for the absent days?." To him, breakdown means he may not be able to perform regular work at the office. He works at MADA as a sweeper. The household income is 15,000 rupees/month. Two of his daughters are married. Mantu, his wife, and his unmarried son Deepak stay together. He is aware of the Chas Hospital and PHCs, their location, and what treatment is provided. However, he does not prefer seeking treatment in these government hospitals. In minor illnesses, even Mantu realises the easiest way is to approach the druggist at the medical shop and take a few tablets (*goli*).

Nuniya Devi, widow of Madhu Hadi, feels ascertained that Doctor Vimal Singh has cured her health problems whenever she visited his clinic. She does request Doctor Vimal Singh to provide some low-cost tablets that can cure her illness. So, there is a perception that cheap medicine should be an effective tablet. She said, "I had a bad cough for over a month and never reported to a doctor. I attempted to cure my cough by taking a few tablets at the medical shop. However, after a month, when my cough persisted for that long and started troubling me during the night, I visited Dr. Vimal Singh's clinic and narrated my illness. I was unable to sleep properly, which forced me to consult the doctor. He administered two *suya* on alternate days to control the cough. A few tablets for one week helped me and cured my chronic cough. I am sure Doctor Vimal Singh can treat most diseases and save our lives."

Nuniya believes strongly, not hearing from her friends or caste members, but it is her own experiences in seeking health care treatment at Doctor Vimal Singh's clinic that is satisfactory. She does not see any barrier in seeking treatment from him. She tells everything to the doctor about her bodily discomfort and feeling unwell. Even her familial problems she discusses. She said, "The Doctor does understand our problem and, in turn, explains properly." Therefore, experiences in the illness treatment process are quite positive and continue to help in maintaining a healthy social relationship between most Hadis and Doctor Vimal Singh.

Doctor Vimal Singh narrates how he has been helping poor patients for the last 20 years in this locality. RMPs are conscious that government does not accept them as trained physicians but the contribution of such RMPs is much more than their status and expertise. Dr. Vimal Singh raises a question to justify his popularity, "*aap hi kahiye, yadi hum log faltu baithe hain to hazaaron logo ka treatment hum kyon karte hian aur hamara value bhi hai yahan ke logon mein. Yadi MBBS itne hote to yahan Bauri Cooli mien kyon nahi aate*," literally you tell me, if I am sitting uselessly here then why thousands of people have been turning to my clinic for over the last 20 years. I am providing care and treatment to my patients, and in turn, they respect and value me. If there are trained MBBS physicians, then why do they not visit and practice in these areas?

Further, he reveals that most trained doctors want a clean and decent (Sundar) area to practice. Vimal Singh viewed the entire location and care-giving process as socially and economically driven. Famous doctors do not wish to meet patients like Bauris and Hadis, who are engaged in dirty occupations. Hence, the entire location of RMP clinics and practices is not only the failure of a health department. But also how caregivers perceive the patients' needs while seeking care during illness. This fact is dependent not only on the kind of earnings of trained physicians but also on the location of the clinics, which makes the difference as they are closer to Hadi Cooli. Hadis do not think consciously about the clinics and evaluation, but it is based on a long-term experience and interaction with the doctors (RMPs) at these clinics that have developed the relationship of Hadi patients with the doctor.

4.6 Stages of Illness Reporting

Ignoring or avoiding visiting government hospitals is an important aspect of Hadis' health behaviour for various reasons. Mantu finds it easier and more convenient to consult Dr. Vimal Singh than going to any hospital. He does not want to be absent from work. Minika Devi, his wife, added during one of the interviews, "*Davaee to thike hau kintu ki dikkat hau woho to sune pare chahiye konho dactor ke. Hame yehe cheezva milo hau docator Mandal theene. Badii baes nukur pucho hai beemariya ke.*" Literally, it is important to hear the narration of the problem carefully and patiently.

I find such an aspect in his treatment at the clinic. Dr. Vimal Singh provides time to listen to the patients."

Another popular consideration in seeking treatment is the availability of a doctor at the clinic. However, there is another doctor, Ganesh, in the vicinity who provides treatment to Hadis. Ganesh has his clinic in Ansari Mohalla, a little away at a distance of less than a kilometre. But Hadis are primarily dependent upon Vimal Singh's clinic. The treatment and illness reporting are usually adopted stepwise by Hadis. But it is never a permanent set of behaviour. Hadis is also found to be shifting between the two, Chas Medical Shop and Vimal Singh's clinic. However, it is to be noted that the frequency of buying *doo goli* (two tablets) is very high at the medical shop. Combiflam and Diclofenac tablets are easily prescribed and sold at the counter in case of any reported body aches and pain. It is a common phenomenon that working Hadis report body aches and pain due to the nature of their work combined with low intake of nutrients. Working Hadis attempt to overcome pain and cure the illness with the help of tablets. Suhail at the medical shop suggests the required medicines, and the same which Hadis usually purchase.

The following framework is developed based on the field data and shows the three levels of encounters and how perceptions get constructed among Hadis relevant to this chapter. This chapter contextualises the following three ways as stages emerged from the data:

First level-Personal Perception of Illness.

↓

Second level-Visiting a nearby Medical Shop for Tablets.

↓

Third level-visit the Local Doctor (RMP) in the Bauri Cooli.

Every day, Hadis are seen visiting Chas medical store for goli in Ansari Mohalla, owned by Aijaj. *Goli* becomes an essential means to control the illness ' Please give me two tablets; I have too much body ache', they commonly ask. The data provided by Jhomdai Devi, a 47-year-old widow of Sipeta Hadi, shows the available methods (here, seeking tablets from the medical stores) to reduce recurring feelings of illness and discomfort from aches. She also suffers from the illness of her husband, who had tuberculosis (her husband's TB and asthmatic problems, as discussed earlier in Chap. 3. But presently, it is not her husband's illness. She herself had certain health problems. She had respiratory problems for some time, similar to her husband. However, some community friends suggested it be perceived as a normal cough. It has been noted from the field that Hadi women fall prey to illnesses related to cough and frequent fever. Jhomdai had a kind of infection that troubled her in breathing, and it was undiagnosed for years. Her husband took her to Dhanbad Patliputra Hospital (attached to the government medical college). Several investigations were conducted as including X-ray, blood, and sputum (hooka) tests. She became unwell. This illness made her steadily incapable of performing everyday domestic work. Jhomdai felt she could not avoid sitting for hours in front of the hearth *(chulha)* to cook for the family. But it was not known whether the same disease had already attacked her husband, Sipeta who was alive till then. They had doubts about having the same disease that panicked her. But she was appropriately diagnosed after all investigations. Cough and respiratory-related problems got prolonged due to delay in the seeking of treatment. In the follow up, Jhomdai continued consultation at Vimal Singh's clinic. She found treatment and follow-ups better.

Jhomdai elaborated on the prolonged illness of her husband that was undiagnosed for a long period. He died in 2007. Jhomdai reported that, "*Khonki badi hovo halae. Raaet bhar kakhnau sute na parto halae, kintu bujhain parau haliye. Bihaane kama par jaa hale to thike raho halae. Voho (sipeta) khato halae ke nae kutch bhael hai.*" Literally, he (her husband Sipeta) had a severe and chronic cough. He would not sleep the entire night. However, we could not understand the pain and the risk to his life. But, during the day, he could go to work every morning; his cough would bother him, but he could attend to his duty. He always wanted to attend work undisturbed." The illustration of Jhomdai does inform how Hadis avoid visiting the hospital or doctor for their initial period of discomfort.

Kamal Hadi, 27-year-old son of Jhomda Devi, said, "Mostly men avoid going to hospital for minor illnesses. It will hamper our work, and half a day's absence is marked in our register." He continues to support that there are certain situations when Hadis avoid doctor's consultations. In his view, the illness of his father due to prolonged breathlessness and persistent dry cough would have been cured if it had been reported to Chas Hospital or BGH. Certain deterring factor minimises visits to big hospitals. Kamal also discussed how it is important to inform and listen to suggestions within the household when in any health problems and sufferings that would help in recovering from the loss of a near and dear. His father did not hear his mother Jhomdai and father in-law who kept telling him to consult doctor. He would have been alive if diagnosed early.

It is important to analyse the fact that chronic illnesses were not perceived as threatening to life. In the case of Sipeta Hadi, who perceived his cough-related problems as a common health issue. The tendency to ignore or seek temporary relief at a medical shop or consulting Doctor Vimal Singh is quite common. There is the possibility of high morbidity that has also minimised the longevity of Hadis. Only a few (4) Hadi have attained the age of 65. Of them three are female and one is male. Perhaps due to undiagnosed and sometimes delayed diagnoses of diseases related to their condition of life, such as TB, asthma, and liver-related problems. Moreover, other habits like alcoholism have impacted their longevity. One understands that they have less information to reason the causes of shorter longevity. It is a relatively significant observation in Hadi Cooli.

However, coming to the issue of Jhomdai Devi, she received medicine from a trained physician at Dhanbad and gradually got cured. In her tone, "*hamni naae dhyan de hiye hau, Jatae doctor jeebin watae aare kharcha karbeen.*" Literally we cannot give proper care to an illness. The expenses rise if we do not minimise visit to the doctor." An attempt to the consultation with a doctor does show Hadis utilising a medical shop instead for minor illnesses. This feature describes Hadis not paying immediate attention to the nature of illness and the way medication is sought in the form of *goli* (tablets). I viewed this as perceiving health care services through *goli*. The patients are not always aware of the medicine, like the names of the tablets and capsules provided. If minor illnesses are not addressed with *goli* (tablets) from *davaee dukaan*, Hadis will turn to RMP clinics. In fact, the majority of the Hadi patients have their first visit to seek a cure either at the davaee dukaan or at RMP clinics.

4.7 Childbirth and RMPs

Now, it is important to understand how Hadis see their health and the factors that affect their health. In the preceding analysis, there are a few factors that come out in the discussion affecting their health; they are open defecation, resource crunch for health care, and distance to public health institutions. There are other problems that challenge the doctor sometimes. In the case of childbirth, Hadis seek the help of a doctor along with the traditional Dai. Dai belongs to the Bauri caste, Shakaro Cooli.

People feel comfortable approaching a nursing home, but it is a costly affair to manage. Therefore, home childbirth is the first preference. During pregnancy, they consult Dr. Vimal Singh at the clinic whenever there are some problems. Doctor Vimal Singh said that the health of pregnant women and also home birth carry some risk to women and the infant. The post-partum period or postnatal period that lasts for a few weeks would have risk due to lack of medical assistance from a trained medical practitioner. The traditional birth attendants (TBA) are not well trained to handle critical conditions, Doctor Ganesh said. As a doctor at Neelam Nursing Hospital and Research Centre (NNHRC) agreed, "Many poor patients do show an inability to bear the cost of institutional delivery through such facilities. There is always a risk to the mother and newborn baby at birth, and a Dai would not be able to handle certain complications, such as bleeding takes place, delay in birth, and even pregnancy pain."

The doctor at the Neelam nursing home informed, "I have seen a few patients who come after delaying a few days in illness. It is quite late to treat poor women and children who would be at an advanced stage during childbirth. It is difficult to save the mother in cases such as stillbirth. It makes it difficult to save the life of mother sometimes." The reason to have home birth is considered a traditional practice among Hadis and it was among other castes also. But there are ANMs and other persons who persuade women to avail institutional delivery. A small amount of Rs. 600 rupees is given as an incentive to the women accepting tubectomy for family planning at BGH and Rs. 150 rupees to the motivator who takes the woman to the hospital. A large number of Hadis prefer to go to their own doctor rather than to hospitals. It is important to observe that the focus of health workers is not on improving and delivering health services during childbirth but on earning a certain amount as a motivator after the procedure is conducted. Hadi women's health is affected due to the lack of a delivery system at Chas PHC. Sonal Devi, aged 18, delivered two children at BGH. They were charged only 100 rupees. ASHA worker who is available primarily at a school in Muslim Mohalla informed that Hadis usually deliver at home.

Interestingly, the point is that firstly, Hadis are found perfectly convincing in the way reasons and causes of illness are deduced and the possible ways of treatment are explained. Second, the druggist is not convinced that his knowledge about medicine is understood, perceived, and accepted by Hadi patients. He is also not bound to give extensive time and explanation for the simple reason that there are no consultation fees asked for. However, Hadis do understand some explanation that brings them to buy medicines at the medical shop. Hadis narrate their conditions of not being well and seek medicines while paying the cost of '*goli*'. Suhail comments, "It is important for me to confine myself to my role and not to enter into the domain of doctor-patient that is witnessed in the clinics." He gives explanations for most diseases and their consequences. This command over certain diseases and the required tablets allows him to suggest the medicine and consequently its acceptability by the Hadis. For instance, Suhail says to a young girl, who sought *goli* for dry cough, "*Khonki baidh gail hau ab goli nae syrup liel partau tabhe thik honbhi–nooni.*" Literally, your cough has increased, and now tablets will not help to control; you need a cough syrup, my dear. This perhaps shows how even a medical shop owner crosses the domain of a trained and skilled doctor regarding medicine prescriptions for the Hadis. Mostly, there is a constant crossover in the boundary of understanding and knowledge between a trained doctor in medicine and a medical shop owner with cursory knowledge about medicines, which they gain through prolonged expenditure in the market of medicine sale.

The medical representatives of pharmaceutical companies who sell medicines to these medical shops have a great role to play in familiarising the medical shop owners about the tablets and medicines. First, the point of crossing over to the boundary of a trained doctor by a medical shop owner shows his ability to suggest the latest medicines for most diseases. The regular interactions with medical representatives (MR) in buying medicines provide ample opportunity to learn. In a situation when a shop does not agree to continue selling a medicine or when new drugs are launched in the market, the MRs try to explain the medicine to the shop owner so that it is bought and kept by shop owners. Suhail and even another shop owner at Bauri Cooli, opposite the clinic of Doctor Vimal Singh, receive MRs from several drug companies weekly. Some MRs have better knowledge about certain medicines, which helps them persuade medical shops to buy and sell them. There is also competition among the MRs in selling their brand. But the shop owner is obliged to do most MRs.

It can be understood from the field data that symptoms compatible with the demands of the person's usual activities, i.e. symptoms which do not cause a hindrance to one's daily activities, are ignored, adapted to, or normalised. The life of Kamal Hadi, 27 years old, illustrates the fact that he has not visited a doctor or hospital for minor illnesses in the past until it is difficult to cope. Kamal recalls that during the winter of 2007, he fell in the Garga River while returning from work. He fractured one of his arms. However, it did not deter him from working with the fractured arm. He consulted the doctor. Hence in his case, many symptoms are responded to in this manner, including those which could be the early stages of chronic disease. In contrast, those symptoms restricting normal activities are likely to be perceived as a problem and presented for medical consultation. Daily life is affected if fallen ill, but the pressure to earn every day gives different dimensions to the illness and work of Hadis.

The sick roles are performed by Hadi patients within households, at *davaee dukaan* and in neighbourhood clinics. A few scholars, like Mechanic (1959), elaborated on the usefulness of the concept of role in the study of illness and disability. It is important to see whether a person is willing to report to health care centres, ready to be absent at work, and agree to be dependent on others' help, then the person enters into the sick role. There is a constant effort by a Hadi worker not to be absent for a day at workplace in order to avoid losing one day's wages. For example, Kamal Hadi, as mentioned earlier, did not want to take a rest and rejected the suggestion to be absent from workplace. In a similar situation, another person taking on the sick role would be ready to go on rest when prescribed by a doctor.

Kamal Hadi himself examined the intensity of his health breakdown, and consequently, he sought medical support to address his fractured arm and later, the doctor advised him to take a few days' rest. But he continued attending to his work. The diagnosis and treatment needed him to rest from work gave him a certification to halt work, and it also gave shape to his mind. But he did not enter into the sick role. Such roles generally demand social sanctity. There is moral and psychological support generated within the households and among the neighbourhood community. But he argued that, "My friends told me it is quite an insignificant fracture. Why should I lose one week of wage? The days will pass quickly, and it is possible to manage the job at the workplace." His household members opined that he should avoid doing heavy work for a few days and say to his co-workers about his pain. By doing this, Kamal could avoid further damage to his fractured arm and not lose his wages. Although the doctors advised them to rest for proper recovery, their financial condition did not allow them to take leave. Hence, he perceived his illness according to the financial conditions in which he lives. Kamal entered a sick role only so far as he got medical treatment for his fractured hand. But not abstaining from work, he did not take sick role entirely.

This is why perception based on the degree of illness and the symptoms affecting one's daily activities gets social acceptance. Consequently, the role performance of a sick person comes into play as a social fact in everyday life. Take an instance, the sick role during pregnancy (used by Rosenberg 1962a, b). He used the Parsonian model to understand the sick role expectations. Pregnant women hold sick role expectation. The role of a pregnant woman is not a sick role, but she may enter into a sick role upon childbirth. This also puts out an argument that a social role is also determined differently at different points in time.

A Hadi woman learns about pregnancy much before she conceives. She knows that pregnancy is a normal occasion in a woman's life. She also knows that the best support is given at home birth. For this reason, an old woman, Pari Devi, says, "What is great about the role of a nursing home in case of pregnancy? All my children were born at home, none of the women were taken to hospital during labour pain, and nursing homes are the places for rich families." For Hadi women, home delivery was a common and ideal method followed by most Hadis. There is also an increase in visiting nursing homes, but only in the cases where there are complications in the pregnancy, women are admitted to a hospital or nursing home. It is also noted that women during pregnancy do not dissociate with the routine activities at home. There is no change in the role performance before and after the pregnancy. Pari Devi delivered all her children at home. She said, "Home birth was most common when I was married. With the help of Dai from Das Cooli on one Western side of the road, most Hadi women gave birth. Pari was attended by a Dai during her pregnancy, and she gave birth to four children; one son could not survive after birth. Two daughters and one boy survived, and today, all of them live happily and are getting married. I did not take rest during pregnancy and kept doing all my household jobs."

The process of discussing the illness within the households and later reporting at *davaee dukaan,* doctor's clinic, and at times visiting BGH (in case of emergencies) is found among Hadis seeking health care services. A few of them visited BGH when the illness was not treated as it is a better-equipped hospital and has good doctors. The other patients visited the Sub-Divisional Hospital (SDH); the district hospital is poorly equipped and has relatively limited medical facilities. The present argument on illness and illness reporting helps bring the processes of health care seeking and the various forms of role performed by a Hadi patient. There are reported role variations of Hadi Cooli based on the age and gender of the Hadi patients in the process of both seeking health care and performing sick roles. Children and aged persons accept the sick role relatively at an earlier stage than a working Hadi. Ashish Hadi, at the age of 61, earns Rs. 2800 rupees by sweeping at the Bank of India office and refuses to take leave in case of minor illness. He has had pain in his thigh for a few years. The pain subsides after taking *suya* from Dr. Vimal Singh. He said, "Why should I take leave for small health problems? I have relatively less work. I need to clean one toilet sweep the floor and do one round of wiping. This needs an hour to do, and then afterwards, I can stay at rest in the office till noon hour." He finds himself struggling against his minor illness and reports at workplace. Several women have also been delaying their illness to be reported in order to save some money. The clinic of Doctor Vimal Singh is one centre to observe the illness reporting pattern and the age wise variations as well. Jadish told, "I have to take *suya* from Doctor Vimal Singh. During winter my pain increases. Early morning is tough to me.

Quite often, the illness breakdown is not considered serious, and adult male members are expected to perform normal activities. The word normal refers to staying physically active and performing everyday activities in order to not lose wages at the workplace. Hence, a sick role exists when a Hadi person fails to report at the workplace, and his physical inability forces him to stay at home. The nature of work demands them to stay fit and report at workplace every day. In this milieu, the role of a Hadi worker depends on his or her performance of his job performance and brings earnings for everyday sustenance.

The existence of sick role performance does not get attention in the everyday lives of Hadis. During an interview with Bajaj Hadi and his wife, Gita Devi, she says, "Who cares about our illness? We carry on with our household chores without any stoppage. But if Bajaj falls ill, it is not acceptable. All of us will suffer. He does not take leave from work. Sickness is not allowed. In case of any illness, I ask him to visit a doctor or get some *davaee* (medicines) at the medical shop at the first instance."

Earlier, it was discussed that the sick role is not perceived as a desirable role to a Hadi, especially among the working members. That close relationship affects earning if not reported at work and sick role. Gita Devi would prefer sending her husband to the doctor for a similar minor illness. She viewed Bajaj attend being physically active and capable of attending to his work.

The degradation of health status due to filthy jobs is one of the reasons that has made Hadis consume alcohol (locally made alcohol) before engaging in menial jobs and after returning from work. It is an important observation to find some Hadis buying alcohol at Bauri Cooli in evening hours. They drink in the same locality or go near *Joria* along with the peer group. The level of consumption has helped to bring down life expectancy of Hadi male workers and resulting in an early death. The illnesses of Hadis were not reported at right time and due to lack of diagnosis, the disease mostly took critical conditions. A few cases of Raju, Jhagu, and Sipeta show that had the diagnosis been done at the early stage, some of the life would have been saved. Somesh, Ashish, Shailender, and Ramesh said that their friends would consume a lot of liquor after returning from work. Those working at Steel Plant offices would consume a lesser amount. The above-mentioned persons would not drink liquor during work hours, but it was opposite to those who worked in other offices and households. Abhishek, aged 62, said, "It is part of our work to consume liquor. Two decades back, almost all houses had dry toilets or open pits. We would go clean it and dispose of it. We had to walk half a kilometre to get far away down to *Joria*. Farmers would not allow us to dispose of their fields. After attending this work, most of us would buy liquor on the way back to Hadi Cooli and consume it liberally."

Doctor Vimal Singh narrates how health is affected due to heavy liquor consumption among Hadis. He says, "The tendency to drink is more after they return from work but not before attending to work. In the evening, Hadis visit Bauri Cooli and buy a pouch of alcohol and consume more than a pouch. They waste money and degrade their health status, too. They could eat good food, but who will tell them? However, there are some positive changes that could be observed; Hadis are less frequent in visiting to buy."

My study uses the framework of Anderson (1995) to understand the use of health care services among my studied population in Chas. Anderson developed a model of health services used in the late 1960s that originally focussed on the family as the unit of analysis, and later, he modified his unit as an individual in the paper titled, 'Revisiting the Behavioural Model and Access to Medical Care: Does it Matter' published in 1995. As discussed in the theoretical framework in Chap. 1, the unit of analysis Anderson has used is a family and how a family utilises the health care services in illnesses. However, his analysis had its own limitations due to the heterogeneity of family members if the family was to be taken as a unit of analysis. Hadis belong to one caste group. All of them do and are engaged in sweeping related jobs either in private offices, Nagarpalika, or steel plant offices. The income level of each household varies depending on the number of earning members in the household. The variation of income level does influence and guide a household's illness behaviour and health care utilisation pattern. Which doctor and hospital shall be chosen in illness is guided by the income level as well. The other aspect is the occupational association of Hadis to sweeping, which has been for generations and also influences and constructs the illness behaviour of a Hadi patient. There is a point when an individual appears important when unwell and sick who is the prime earner of the household, then other household members, and so on.

Although the medical care an individual receives is a function of the entire family, it is also important to specifically look at the individual as a unit of analysis. This understanding provides the basis for the present research to study Hadi individuals as the unit of analysis and how they access health care during illnesses. The use of medical care is located in the individual situated in the household and the importance attached to individuals. The utilisation of health care is governed by various factors within the household and external factors outside the household. For instance, the importance of an individual is decided within the house, which is the head of the household, preference to soft populations like children and old people are given preference over women and young who are understood as capable of coping with illness. The older Hadis who are working people manage their health problem at Davaee Dukaan. Shailender and Abhishek, at 62 years old, work as sweepers in private offices. Mostly working Hadis avoid taking treatment that a few goli can manage.

The discussion within a household about an illness takes place. Mantu is described as "I should take care of my young children. They should not be left without proper treatment until coming out of illness. Children are not able to explain the illness and the intensity of health breakdown, unlike elders like me. I can tell the doctor about my symptoms and all the feelings and problems I have faced of being unwell. But I saw my daughter when she was 6 years of age. We could not understand her ordeal. Sonam had indigestion and problems passing urine, but she always said her inability to eat anything but could not say exactly the problem lay in her urinary infection. It took one month of treatment process and later diagnosed her infection." Sometimes, children do not explain their problems and are ignored as minor illnesses; however, it may lead to a major illness if ignored. Sonam's investigations of blood and urine diagnosed the disease, and follow-up medicine cured her in a week. Thus treatments for these populations are either ignored or postponed temporarily. This maximises the household expenses on health care.

The preference to seek and provide health care is generally not pre-decided in the Hadi households instead they are spontaneous. The decisions to provide health care are decided according to the intensity of the illness perceived and accordingly ill person is provided necessary treatment. Hence, I took the help of Anderson's model to understand the perception of health and access to health care among Hadis with the field data, and they are elaborated in the subsequent Chap. 5. For instance, Ronald M Anderson argues that community and personal enabling resources must be present for use. The first instance first, "Health personnel and facilities must be available where people live and work, and secondly, people must have the means and know-how to get those services and make use of them" (Anderson 1995, p. 3).

He gave a description of predisposing characteristics in his model of health care utilisation. The predisposing characteristics comprise four components, namely:

firstly, demographic (age and gender) and health beliefs; secondly, enabling resources such as personal or family and community; thirdly, need as perceived; and fourth, use of health services as important factors that play a role (Anderson 1995). The role of Hadi households is important and reflected in the decision-making process for seeking health care services. Age and the importance of the person in the household are a few considerations that help to decide preference. Households (for individuals) play an enabling resource along with the predisposing factors to provide all the necessary support in the forms of taking to the clinic, bearing the expenses of treatment, etc. The need to use health services is mostly individual, but it is mostly a matter of the household. A few *goli* is sought at *davaee dukaan* individually at times, but they are also bought by any member of the household members by sending another person. The connection among the above four factors is important in Hadis's life. I studied Anderson's health care utilisation model partially and took components of the model and not the entire model.

Further, the understanding of enabling factors and putting those factors to Hadis' health care behaviour and illnesses makes it essential to examine the perceptions of Hadis and the way they seek health care at medical shops in RMP clinics and later at hospitals. The revised model helped me understand why not family but how individuals use health services. I see it quite the importance of the use of health care when community and personal enabling resources are present. Know-how and means, as Anderson writes in his model, are quintessential to people having them. I find it relevant to Hadis' health care utilisation practices. They avoid visiting and consulting doctors at government hospitals and visit Davaee Dukaan and Dr. Vimal Singh's clinic. This health behaviour reflects their know-how about available health care in town, but other factors of Anderson's model are also important. Income, travel or commutation, and waiting times at the Chas hospital do deter them from visiting in illness. They perceive health care as effective at the clinic of Doctor Vimal Singh and Davaee Dukaan is accessible. Waiting time, consultation fees, commutation, and travel time do not create hindrances to the-counter medicine (OTCM) at Davaee Dukaan and at their preferred doctor, Vimal Singh's clinic.

Factors which minimise access to health care centres depend on the fact that one, the location of health care centres acts as a physical barrier and two, the whole process of accessing health care centres and providers where their cultural status acts as a significant social barrier. Nuniya Devi, who lives with her younger son Jhomda Hadi, had experience visiting the Primary Health Centre (PHC) located at the check post area of Chas town. She is an old widow, aged 65 years and finds it quite difficult to visit the PHC to seek treatment for her illness.

While coming back to Hadi Cooli after her revisit to PHC in the first week of December 2010, she said, "During winter, my hands and legs become so painful that I become immobile. However, I take care of most of my personal needs. It is only my courage that I not only manage alone but also take care of the daily chores like cleaning utensils, sometimes sweeping the house and the veranda, and taking care of children's needs. Otherwise, it is like falling down due to weakness, and I would become a burden on my daughter-in-law. Today, I am feeling a little better after meeting the doctor. I rarely visit this hospital, and why do they visit when they do not give medicine despite it being a government hospital? However, today, I managed to get medicine from the doctor after making repeated requests to him. See these medicines he has given me without paying any money. Two of these have to be swallowed, and the third one is for massaging in the painful areas. Today, two doctors were sitting in the chamber. One of them asked me to show my hands, but he did not bother to touch and inspect properly. Nevertheless, yes they paid attention to my illness description properly. Last week, when I visited, I could not meet the doctor as he told me it was time for him to go home and asked those of us waiting in the queue to come another time. My son does not earn enough, and for this reason, I do not ask him to support me much for my health expenses. He already struggles. He goes to office work early in the morning and returns late in the evening. His office is located in the city centre in Bokaro Steel City. What should we do with an old lady like me? We also do not get good, nutritious food. We also do not get good nutritious food, neither milk nor meat and rice. They are very costly items. Poor households like ours cannot afford such costly food items. Jhomda, my son, brings delicious biscuits from the city centre. It is superior in taste. Manik's grocery shop does not keep this biscuit. What can I eat as good food? Whatever is cooked is served to all the household members. When I think deeply, I realise that I have so many health problems. You are saying that free treatment is available for the poor at government hospitals, but they are just words of your mouth. We will never get free health care services."

The financial capacity of the Hadis allows them, at best, to visit only the PHC, but more often than not, the services are not available. The financial condition of the household creates a dynamic and complex situation among the family members. On many occasions, a constant effort is present among family members to ignore minor illnesses, particularly by adults and older members of the family.

Among the Hadi community, both community, Hadis and personal enabling resources (for example, income, ethnicity), in the sense Anderson must be present to take benefit, and only then utilisation of health care services is possible. For example, the argument by Anderson about health personnel and facilities, mentioned earlier, are two enabling factors that are crucial resources and facilities in the everyday health care needs of Hadis. The caste and income of Hadis play an important role to enable them utilising health care to their perceived illness. The data collected during a couple of interviews shows that an important role in enabling them to utilise health care for Nuniya Devi discussed above, would prefer to make use of public health care if the services are available and provided at right time when she visits PHC during illness. The enabling factor would also refer to the health care services facilities to the people. The location of health care centres, if far away, also shows a disabling factor to patients who are already poor and less aware of the administrative functioning in the health care services. These are available at the low cost that Hadis can afford. To illustrate, let us see even Chas town has two public health centres that are underutilised by Hadis due to certain weaker enabling factors: low income, and menial occupation. The utilisation of public hospitals would increase if it was near the Hadis residential area, easy to travel, or within walking distance, which could help in saving costs on commutation. Therefore, at times, the enabling factor does not favour Hadis, and they seek treatment at the *davaee dukaan* and consult Doctor Vimal Singh at his clinic. The dependency on *suya* and *goli* emerges as an important fact due to weaker enabling factors in the health care utilisation process; Nuniya is one of them that could not be utilise.

There is some relationship between the perception of the people towards health care services and the process of illness experiences in seeking treatment at the clinic in the locality and buying *goli* over the counter of the medical shop. Over time, those experiences are translated and constructed into the perception of the community. The following framework of Anderson attempts to illustrate the relationship between primary reasons for ill health and population characteristics. For example, the socio-economic condition of the Hadi population, housing conditions, caste background, and occupational engagement influence the process of health care seeking behaviour and access to health care services during illness.

I discuss the population characteristics in Chap. 2 of the dissertation, which provides the income and occupation, age, and gender description of the Hadis. In this section, I find the health behaviour of Hadis through three stages, as discussed earlier in this chapter. Hadis perceive their illness in the process of seeking treatment at *Davaee Dukaan and Doctor* Vimal Singh's clinic. The effort to cope with small health challenges, firstly with the help of traditional remedies, is important to Hadis. Hadis access *suya-goli,* either at *Davaee Dukaan or in Dr Vimal Singh's clinic, reflects* Anderson's model partially. It takes place when a Hadi perceives his or her illness and then reports to a doctor or *davaee dukaan*. Hadis make use of *davaee dukaan* quite frequently and component finds comfortable reporting at Vimal Singh's clinic, which shows the use of health services as one of the components of Anderson.

Sukhicharan Hadi's feeling of weakness and pain is not due to his occupation; at 60, he is not as brisk at his work in that office. He works at ICICI Bank as a sweeper. The increased burden on health expenses has affected his savings and decreased his working capacity. As a father, he continues to take care of his household expenses and his wife. The couple lives by themselves. His two sons are married and live separately. He does not go to a clinic for minor illnesses like colds, body aches, and headaches, etc. He instead chooses the medical shop to curtail expenditures and get a few tablets. Sukhicharan said, "Most medical shop owners give you a few *goli* if a patient asks. It is always difficult to cope with illness for longer days as it would affect the work and earnings." His wife, Parvati Devi, is now 55 years old. But she still finds that her husband is very active. In case he stops going to work, his health will degrade gradually. She encourages him to continue working for some time, which will keep him active and healthy. They know that the doctor is available at Bauri Cooli in an emergency. Doctor Vimal Singh can be approached at his house in Shakaro Cooli.

The way people perceive their feelings of bodily discomfort, suffering in illness and the level of pain felt, and various ways to cope and manage such discomforts and pain are all somewhere influenced by the culture of the society. Hadis don't need to establish a causal relationship for each illness they fall. Hadis do not associate a direct relationship with the surroundings and hygiene in their localities in the form of personal hygiene, diet, work, housing structure, and exercise, as well as safety practices and illness avoidance practices. However, a few illnesses are correlated with certain causes. Why Ganesh Hadi sought *suya-goli*, and not treatment at BGH is faith, and not treatment at BGH is faith in his treatment that brought him to consult Vimal Singh. The cultural aspects between the doctor and a patient are an important dimension of health perception and illness behaviour that rely upon the suya-goli of the doctor in the locality.

There is also another facet of this relationship between a patient and the doctor that also depends on the individual's satisfaction with the treatment sought and provided. Some individuals find it quite satisfactory to the treatment provided by the doctor in their neighbourhood and show high trust. Therefore, individual experiences and personality may have an impact on perception and behaviour related to self-care and health practice, but broad cultural association with the health provider plays a pivotal role in accessing the same. Just as each culture has its explanations of illness and healing, each culture also has a health care specialist who provides treatment when self-care[1] fails. These specialists have specific attributes and defined levels of status within the culture to determine what causes sickness and how sickness is demonstrated (Chaudhuri 1990).

Health cultures are an integral component of the overall culture of the community. The culture of a community also indirectly influences its health culture because certain cultural practices such as food and drinking habits, pregnancy and childbirth practices are directly related to the health problems in any of the communities studied (Sahu 1986; Bir 2001). Thus, Hadis live their life with the help of traditional healers and Baba, etc., to cope with and manage health care problems. It is also observed that Hadis give little stress on health care unless their health is adversely affected, but this variation depends on the individual who falls ill.

One broadly understands that health care is perceived through cultural traits learned in the community where one lives and grows. Hadis have developed their own mechanism to manage illness and accordingly seek health care services in order to prevent themselves from falling into bed. They are found to be gradually moving or, to an extent, moved away from traditional healings to the stage where allopathic and other methods of treatment are also followed depending upon the nature and time of illness.

While understanding the health-seeking behaviour of Hadis and their way of perceiving health care services through suya-goli, it is important to understand that most cultures work in a meaningful fashion in their treatment of disease. Similarly, every social group in a given culture evolves its own system of medicine, and the treatment varies from one group to another. The process of evolution in the form of community medicine has evolved differently among different communities over the period. It is understood that most health systems develop to fulfil the culturally perceived needs of health care services in a community, and the same health care system emerges as an important way to cope with most health breakdowns (1990). Illness perceived is a fact, but the way illness perceptions are constructed and shaped

[1] "Self-care is a deliberate action that individuals, family members, and the community should engage in to maintain good health. Ability to perform self-care varies according to many social determinants and health conditions" (World Health Organization 2009, p. 1).

is a long process of interaction and encounters between, for example, patients and Hadi RMPS as well as *davaee dukaan*. The role of *suya-goli* emerges as a significant means to seek treatment and, consequently, Hadis perceived notion about illness and health.

The way people perceive their feelings of bodily discomfort, suffering in illness and the level of pain felt, and various ways to cope and manage such discomforts and pain are all somewhere influenced by the culture of the society.

The urge to quickly overcome an illness is seen among Hadis who go to the medical shop or RMPs. It is noted that *suya* (injections) and *goli* (tablets) help them to achieve this by making them recover from most illnesses much faster and it is less expensive. The interchangeable visits to the medical shop and clinic of Dr. Mandal are found in the majority of minor illnesses among Hadis. However, the satisfaction of being treated and cured is attained largely at the clinic. Because he is a doctor and can diagnose much better than Suhail. The doctor is perceived to be a trained person with much more knowledge than Suhail, who only owns a medical shop. There is some differentiation made between Suhail and doctor for Hadis.

For example, the doctor administers injections and provides bandages on wounds or cuts. He also attends child delivery at home and provides medical help to Dai (birth attendant). He administers *suya* whenever patients are in severe trouble. Injections administered to Hadi patients provide relief to them. The patients also take a few tablets. Hadis have their own way of distinguishing between kinds of treatments and medical practitioners, as mentioned above. The visit to the medical shop is observed as an attempt to address the minor illnesses that disturb the everyday routine and work of Hadis, namely, attending duty at workplaces, household maid jobs, and doing jobs at private offices such as banks, private offices, etc. The sweeping and cleaning remain their main work wherever Hadis work.

Hadis consults the doctor for several illnesses, and they know him to be more appropriate for the issue than buying medicine over the counter. The most severe health problems are noted as vomiting, skin infections, allergies, asthma, tuberculosis, and diarrhoea. But fever, cough, body aches, pain, et cetera, are commonly reported illnesses at medical shops and to Doctor Mandal. Dr. Mandal has to attend to a large number of patients infected with viral fever.

Absence of safe drinking water, unhygienic living conditions, congested housing, poor public health services, and ineffective coverage of national health and nutrition services have been important factors that also affect in increasing the illness of Hadis. All these forms of treatment-seeking aspects are socially defined and also governed by the interpersonal relationship between the doctor and the patient. But they also depend on the financial burden of seeking any treatment. It was reported during a discussion with Doctor Mandal and Doctor A K Singh that illness among Hadis is increasing not only due to seasonal changes and occupational hazards but also due to a lack of proper sanitation facilities, poor housing conditions, and low ventilation. Dr. Singh is a practitioner at Muslim Mohalla, opposite Chas medical shop (shown in map 1). Hadis do not visit his clinics. The filth and dirty drains are a threat to most people across the Mohallas, but Hadis are at higher risk as their houses are much closer to open drains. Most women clean their utensils outside the door. The filth cannot be controlled unless Hadis have cooking space. Air and water pollution are also increasing every year. The data provided by doctors correlates to the higher risk faced by Hadis than others.

It is not based on the period and duration of the treatment but on the way it is followed in the events of Hadis at Dr. Mandal's clinic. The encounter with a Hadi patient at his clinic to study illness and health perceptions of Hadis is significant in my research. Hadis is aware that the doctor in the clinic does not have any assistance from either a male or female nurse. There is no barrier or third person at the clinic of Doctor Mandal, unlike any other doctor and hospital outside the locality, reported Lamti. She said, "I visit his clinic. If no other patients are waiting, he would immediately ask after my health and inquire about the problem. There is no counter and appointment arranging person who would issue a consultation slip. There is direct communication with the doctor, and he prescribes according to the illness description given by patients. I have seen so many procedures to be followed when visiting BGH. We get panicked and troubled during the consultation process at BGH and even at Chas Hospital." Therefore, the first interaction is not with a person at a counter as found in other clinics and hospitals in the town. Hadis need not pay consultation fee in advance to the receptionist/assistant and even then asked to wait in queue to meet the doctor. At the doctor Mandal's clinic, it is different from most hospitals and clinics in Chas.

The strong trust established between the Hadi patients and doctors as healers works as the most significant means to understand the various dimensions of illness perceptions among Hadis. Samina Devi once said, "Doctor Mandal is like our own person. He understands the problems and empathises with our health breakdown. It is important for me that a doctor provides some space to a patient that also helps build trust. I have faith in his treatment. He explains the tablets, why an injection is required—if to be administered, and a rough idea of the cost of treatment. My narration is usually not heard in most hospitals I visited, unlike Doctor Mandal." A patient to draw comparisons with the hospitals and doctors in the town with Doctor Mandal in terms of time spent during the consultation, hearing the illness problems, discussing the cost of treatment, etc. These descriptions helped to construct and shape the perceptions of Hadis towards health care as well as such centres.

The discussion at the clinic of Doctor Mandal provided reasons for the frequent illnesses of Hadis. They said viral and bacterial infection poses a higher threat to Hadis. He also highlighted the adverse effects of open defecation during monsoon. It risks increase in gastrointestinal problems in Hadi Cooli. One can observe a large number of patients having stomach-related problems. Except for three households, the rest have no toilets in the house. Somesh said, "People do not understand this as a problem. It has never been an issue and challenge to be addressed. But I see that defection in open areas is not good. Almost all other castes have stopped going out except my people. We must learn from other communities and think seriously about saving some money at the household level and constructing at home."

However, a few Hadis said open defecation is believed to be an age-old practice, so why is it bad? It is quite possible that a few Hadis who have worked in BSL and have seen the residential quarters attached with toilet, and they wanted to have toilets at home. The discussion occurred when the question was raised about why Hadis do not have toilets in their houses. Water availability is very closely attached to toilets and sanitation facility that is minimal in Hadi Cooli.

One of the doctors, (RMP) Mohammad Jamal, in Ansari Mohalla, another neighbouring locality, reports that 'Hadis still wash their utensils and clothes in nearby ponds during summer when the water crisis is at its peak. Dr. Jamal practices at Ansari Mohalla. The pond water is highly contaminated and impure. Hadi Cooli people mind using such water, which other people find impossible to use. Most ponds are used to wash clothes and bathe animals (especially buffaloes) among the four. Even a few households' drain water is channelled into the pond. It is a risk to the health of Hadis who use ponds for domestic purposes.

One could see the pots and utensils being washed in a pond located on the backside to Hari Mandir market. Jamal says, "It is polluted in both senses; water is impure due to contamination, and the second one cannot feel spiritually (so to offer the prayer) fresh after taking a bath with that water." He further reminds us that due to the increased filth in such ponds, women are going every week up to three kilometres distance during summer to wash their clothes." No upper caste uses water from the ponds except travellers like truck and bus drivers halting overnight in Chas. Hence, the non-existence of toilets adversely affects the health of people who are living close to this filth.

The causes of illness during monsoon are known among the medicine practitioners in the vicinity due to open defecation, contaminated drinking water, and exposure to filth at work. The attempt of Hadis to seek *suya-goli* at the doctor's clinic is a medical condition of the patients when they attempt to look for treatment and get cured. For them, Ibrufen and Paracetamol are magic medicines. The clinic provides all forms of treatment at the first instance whenever a patient visits. However, patients are also referred to Chas Hospital when it is beyond the doctors' understanding. Hadis do follows his suggestion and goes for a follow up. Doctor Mandal said, "It is a free consultation and sometimes several investigations are conducted at a very low cost if the equipment is functioning." Therefore, suya-goli is perceived as a mechanism for managing illness among households in the Hadi Cooli. This provides the method to study Hadis' perceptions to illness and health through *suya-goli.*

References

Anand, K., C. S. Pandav, and S. K. Kapoor. 2001. Injection Use in a village in North India. *National Medical Journal of India.* 14 (3): 143-4.

Anderson, Ronald M. 1995. Revisiting the Behavioural Model and Access to Medical Care: Does It Matter. *Journal of Health and Social Behaviour.* 36 (1): 1-10.

Bir, Thaneswar. 2001. *Dynamics of Health Culture: Urban Slum and Behaviour.* New Delhi: Rajat Publications.

Chaudhuri, B. 1990. *Cultural and Environmental Dimensions of Health.* New Delhi: Inter-India Publications.

Garg, C. C., and Karan, A. K. 2009. Reducing out-of-pocket expenditures to reduce poverty: A disaggregated analysis at rural–urban and state level in India. *Health Policy and Planning, 24*(2), 116–128. https://doi.org/10.1093/heapol/czn046

Government of India (GoI). 2014. National health profile 2014. Central Bureau of Health Intelligence, Ministry of Health and Family Welfare, Government of India.

Kermode, Michelle. and Vanlal Muani. 2006. Injection Practices in the Formal and Informal Healthcare Sectors in Rural North India. *Indian Journal Medical Research.* 124 (November): 513-520.

Kotwal A. *et al.* 2004. Injection Practices in a Metropolis of North India: Perceptions, Determinants and Issues of Safety. *Indian Journal of Medial Science.* 58 (8): 334-44.

Mahapatro, Meerambika and A. K. Kalla. 2000. Health Seeking Behaviour in a Tribal Setting. *Health and Population- Perspectives and Issues.* 23 (4): 160-169.

Reeler, A. V. 2000. Anthropological Perspectives on Injections: A Review. *Bulletin World Health Organization.* 78 (1): 135-43.

Sahu, S. K. 1986. Social Dimension of Health of Tribals in India: A Case Study of Oraons of Orissa. In Buddhadeb Chaudhuri (ed.) *Tribal Health: Socio-cultural Dimensions*, New Delhi: Inter-India Publications.

Shahrawat, R. and K. D. Rao. 2011. Insured Yet Vulnerable: Out-of-Pocket Payments and India's Poor. *Health Policy Plan.* 27 (3): 213-21.

Suchman, Edward A. 1965. Stages of Illness and Medical Care. *Journal of Health and Human Behaviour.* 6 (3): 114-28.

van Doorslaer, E., O'Donnell, O., *et al.* 2007. Catastrophic payments for health care in Asia. *Health Economics, 16*(11), 1159–1184. https://doi.org/10.1002/hec.1209

World Health Organization. 2009. *Self-care in the Context of Primary Health Care: Report of the Regional Consultation Bangkok.* Thailand 7–9, January 1-2.

Zola, Irving K. 1973. Pathways to the Doctor - from Person to Patient. *Social Science and Medicine.* 7 (9): 677-689.

Chapter 5
Experiencing Health Care

5.1 Introduction

This chapter discusses the multiple factors influencing Hadis' perceptions about cost management in accessing different types of health institutions for health care. The higher cost of consultation and treatment in private health care institutions is a barrier to accessing health care. A discussion on private and public health care also brings to the surface the process of dealing with the illness of Hadis. The increasing cost of treatment in urban private hospitals and clinics run by qualified doctors is out of reach for patients like the Hadis. This chapter analyses the experiences of Hadis with various health care institutions in different types of hospitals, RMPs, and traditional healers.

I have described utilisation of health care services by Hadis. The health care-seeking experiences of Hadis at different sets of hospitals like public health centres, private hospitals or clinics (having qualified doctors), traditional healers and RMPs' clinics in the vicinity are inter-connected. I attempt to further explore the various reasons that make Hadis explore different health practitioners in the course of seeking treatment.

5.2 Experiences at Public Health Institutions

Chapter 2 discusses how public health institutions are meant for all the citizens of the town, especially those crucial to poor population groups like Hadis. The population increase in the last 10 years has also burdened the patient load in Chas Hospital, one of the public hospitals in that area. The public hospital is not popular with the Hadis while accessing health care institutions, though they prefer accessing public health centres.

K. M. Ziyauddin, *Occupational Caste Groups' Illness and Health in India*, People, Cultures and Societies: Exploring and Documenting Diversities,
https://doi.org/10.1007/978-981-96-8967-5_5

In Chap. 4, the unfriendly and time-consuming functioning of government hospitals is identified as the primary reason that guides Hadis towards RMPs and purchasing medicines over the counter. This pattern creates impediments in regaining good health for Hadis as they seek medical care from a non-qualified doctor. The trust in the health provider is also largely dependent on the quality of interaction between the doctor and a patient. Hadis has formed and constructed their perceptions about the functioning of public health care facilities based on their past experiences, and these are later shared among others in the Cooli. The experiences are generally unpleasant experiences. The common fact emerged among Hadis that free treatment is good for saving money only if treatment is readily provided. They have high expectations, especially from public health institutions as a centre providing free medical facilities. According to them, being poor and a member of a low caste should entitle them to get free medicines. However, these expectations of Hadis are not fulfilled at the public health institution. In my observation of Hadis, it is a big step to approach Chas Hospital as one visit to the hospital implies losing one day's wage.

Hadis have a range of expectations that are not met. For example, Hadis finds the consultation timings problematic due to the conflict between doctors' availability and the working hours of Hadis. The patient department (OPD) is open from 9 AM to 1 PM. Most working Hadis have to attend their work in the morning hours. They usually return home by afternoon. The incompatible timings of the two hospital's OPD and Hadis duty hours hinder them from accessing public health care. It is a deterrent to them from availing themselves of public health care services. Visiting a medical shop and clinic of Dr. Vimal Singh does not take several hours. One can seek or buy medicines within a few minutes if a clinic or shop is open. Hadis complain that doctors in public health care centres are available for a very short period. It is the hospital administration that has fixed up the shorter consultation timings, which restrict interaction between the doctors and the patients.

Additionally, on several occasions, public health care institution doctors were not available in the hospital during the scheduled consultation hours. Others in town reported that hospital doctors are more inclined to do private practice. It is believed that one of the reasons that doctors are not willing to do consultations is that they are said to be doing private practice. Though government-employed doctors are not permitted to practice privately, a few doctors are said to have some sort of understanding with their sub-ordinate staff, like compounders and the person issuing consultation slips, which facilitates them in running their private practice. In return, the compounder finds it easier to be away from work. There have been instances when doctors at the government hospital were given show cause notice as told by a civil surgeon's office staff. It is commonly observed by the people in the town who commented that the doctors were practising outside and staying away from the hospital during their working hours, while people find it time-consuming while wait for hours to consult a doctor in a government hospital.

Due to all this, Hadis experienced the non-availability of doctors, medicines, and investigation facilities in the government hospital often in their limited number of visits. They lack access to doctors, medicines, and investigation facilities. According to them, visiting Chas Hospital is futile when treatment is unavailable.

Nevertheless, in certain situations, Hadis sought medical care at PHC and Chas hospital. Dulari Devi, aged 60, visited the PHC several times. She has to reach the hospital on foot to seek treatment despite her old age. She is widow of Jamal Hadi. She said, "*Hamni ke thik lago halae vahan. Jadvan mein kakhno chael ja haliye check postva. Abe hamae nae jae parbo. Jaldi jae de hathin davaee kintu baithe ke baaed milye jeeto kakhno to.*" Literally, we used to visit earlier along with my husband. I visited the hospital last winter. Now I cannot walk a long distance. Sometimes, I was given medicines as well. Now, hospital administration does not provide medicines. Dulari Devi recalled visiting the PHC occasionally when her husband, Jamal Hadi, was alive. In 2011, she walked to PHC, 3 km away. She finds it tiring to visit PHC for some free medicines. She said, "I reached the hospital quite early in the morning. I did not walk on foot for three kilometres at my age, but I managed to reach it somehow. I was the first person in the hospital. I planned to meet a doctor for my prolonged pain in my joints for two to three years. It increases during winter and reduces once winter passes." She was prescribed two tablets and an ointment to rub on her knees. Dulari requested the doctor to provide the prescribed medicine to her. The doctor refused to provide tablets to her but she did not buzz. After 2 h, when the doctor came out of the chamber, he realised Dulari was still waiting. The doctor was surprised to see Dulari waiting outside his chamber for almost 2 h. After having seen her, doctor called her in the consultation room and prescribed some other medicines. She thanked the doctor for this gesture. But Dulari also recalls another visit when she had returned without medicines few years back. The intervention of doctor helped her to receive medicines. She felt the problems of non-availability of medicines are not genuine rather the doctor and staff do not give. They do not provide medicines to poor. A few other patients are given medicines.

Ganesh Hadi, aged 40, told me, "*Haath-gaud ujjar niyar bhae gail hai kahon kahon, badi surssura halae. Konho cheez ghus gael halae hau.*" (My hands and legs have developed white spots and rashes. A constant skin irritation persisted). Ganesh had developed some infections in both his hands. The Doctor diagnosed some skin infection for which he was asked to buy medicines from outside. They pick up waste materials and garbage that may be infectious. The doctor told Ganesh to avoid going to work for 2 days. It will delay the cure if he works along with the medication. He wondered how he could remain absent from work even for a day. Ganesh did not find the treatment to his satisfaction at the public hospital; besides, no medicines were provided at the hospital. Therefore, he returned to Dr. Vimal Singh's clinic for further treatment. He has no problem in visiting PHC if proper relief is given. His duty is near the hospital. The behaviour of the hospital staff is rude to poor patients like him. Mostly, they are provided no free. When people request at the drug store counter, mostly the store in-charge refuses to provide. The inability to pay for private qualified consultation takes him to a government hospital. If poor patients like Hadis do not get care, it negatively impacts on the mind of Hadis as a patient.

During my initial fieldwork period, a PHC doctor made me wait an hour. Although the nurse had passed on my message to him in his chamber, he did not respond for an hour. Then I went to enquire in his chamber, where he was gossiping with a person over a cup of tea. He asked me to take a seat. I explained to him my research work. The doctor was not comfortable with my purpose of research and tried to avoid talking about the questions I raised. Interestingly, when he realised that I came from Delhi, he started asking me about the admission process for undergraduate students at the University of Delhi for his elder son. This gave me an edge to prevail upon him and I felt better placed. It helped both of us to develop a rapport. He offered tea and *samosa* brought from a nearby sweetshop. We spoke over an hour. The doctor believes that SCs' poor health is largely due to their lack of concern for their own well-being. He attributes their recurring health problems to living in congested residential areas, engaging in uncontrolled alcoholism, and exposure to filth after work. Though he did not visit Hadi Cooli, he heard about Hadis living in Hadi Cooli from his staff working at the health centre. I queried if Dalits do not get equal attention in accessing health care services. The doctor objected to my observation. He said, "Whenever people from lower castes like Hadis and Bauris come, they keep pleading for free medicines. PHCs have limited medicines that can be provided to the poorest. Only Hadis are not poor. We have a large number of patients who are poor and seek medicines. When free medicines are not given, they feel bad about it." In between, he requested that I not take notes during the interview, to which I agreed. We also discussed about the occupational hazards among Dalits.

It is commonly understood that lower socio-economic status (SES) is associated with and assessed by parameters like lower education, income, and occupation, poverty, and poor health of a population. Lower socio-economic status did help to understand the various factors contributing to a lower level of health, various illnesses, mortality, etc. (Adler and Newman 2002), as I discussed earlier in Chap. 2.

The utilisation of low levels of health care in other studies shows the association with discouraging experiences of poor and lower caste groups in India in comparison to non-Dalits (Kulkarni and Baraik 2003). Seventy per cent of households in the Jharkhand state use services of private medical services that include qualified doctors, private hospitals, and clinics for any treatment (NFHS, 2005–2006). As reported by Hadi's, poor access, under-staffed hospitals, low-quality infrastructure, and overall behaviour of health personnel are some of the responsible reasons for the under-utilisation of public health services.

Steel plant employees get proper leave benefits (they are entitled to medical and casual leave). This enables them to avail leave without any deduction in salary. Ganesh Hadi compared BSL with health perks from other employers, "We have regularly available jobs in Nagarpalika. But there is no health facility exclusively for its employees, no promotion, and no assurance of future benefits, unlike Goli Ka (uncle), who, even after retirement from BSL, is entitled to medical care. Any health expenditure burdens us and the government hospital does not provide proper health services."

Poor people like Hadis have high expectations from the government hospital, which do not match the availability of care facilities. Even NFHS 2005–2006 reported lower quality of health care infrastructure in the state of Jharkhand. Under-staffed hospitals and unofficial absenteeism of doctors increase the burden on a hospital. All such problems leave unpleasant experiences when visiting government hospitals (NFHS, 2005–2006).

Pan Babu, 55 years old, said, "Chas Hospital could have been the best place for Hadis to access health care services, but most staff and doctors behave rudely with Hadi. I have seen this a few years back when I wanted to consult a doctor, the receptionist at the counter did not issue a consultation slip to me. I worked 10 years back at the Chas Block Office near Chas Hospital. I recall some official discussing Chas hospital and experiencing satisfactory treatment there. When I visited, I thought it would be easier to avail consultation for my fever that was troubling me for a few days. He (the receptionist) asked me to wait until the doctor took his seat. The consultation time was till 1:30 PM. Almost 10–15 patients were already waiting outside the doctor's chamber with a slip in their hands. I did not get a slip. When I asked again for a consultation slip, I was denied. After waiting for 2 h, the compounder informed us that the doctor was not available that day, so no more consultation slips would be issued. The consultation timings were already over when it was announced. Most patients were sent to another doctor's chamber. I was denied a slip. The doctor did not allow me to consult nor did he hear my plea. I tried telling him why I did not get a slip. The person at the counter had already closed his room, and I was left helpless. For small reasons, we are harassed, which makes many of us feel sad, and we don't prefer visiting Chas hospital." These experiences offer an explanation that even if Hadis somehow managed to visit a public hospital after giving up on the day's wages and spending a few rupees on transport to reach a public hospital, the staff at the hospital created certain barriers in accessing the services. It is hurtful for Hadis in more than one way to undergo such an experience.

The paper titled 'Caste and health in Indian scenario' Vijayanath et al. (2010) analyses the utilisation of health care, poor facilities, social determinants to health care, and barriers to health care services. In their paper, the analysis relies upon the National Family Health Survey (NFHS) 1998–1999 and 2005–2006 data and argues that socio-economic factors are more important determinants of health care than just medical support. They illustrated that lower castes (SCs and STs) have poorer utilisation of health care than other social groups due to factors like caste and income. The problems of "Unequal distribution of power, income, goods and services, the restricted access of those from the lower castes to clean water, sanitation, nutrition, housing, education, health care and employment is due to a toxic combination of poor social policies and programmes, unfair economic arrangement and bad politics" (Vijayanath et al. 2010, pp. 42–43). In contrast to Hadis, who have not been able to access public health care services due to various reasons, including health care provider's behaviour and disparity in the provisioning of health care. The experiences of treatment and discrimination at public health centres have also been a common discussion among Hadis. The low-income groups are more vulnerable, and they have a higher level of reporting early hospitalisation usage like Hadi. Hadis have not been able to utilise hospitals in the town. But low-income households saw higher hospitalisation due to their vulnerability in contrast to higher-income areas where outpatient care was more readily available (Billings et al. 1993). Not only income, occupationally lower people also face denial of public health services (George 2015; CEHAT 1995).

Let me illustrate the case of Nirmala Devi's husband, who could not get proper treatment in the government hospital. Nirmali said, "I made two attempts to show my husband in Chas hospital, but nothing happened. The Doctor examined him and prescribed a few investigations and medicines. There is the detailed explanation given to patients so that we could understand, but the hospital had no medicines and no medical investigations. Then why do we go there? The poor have no care and support even in' government hospitals; who will hear our plea and provide low-cost treatment?"

The experiences at Chas Hospital did not prove to be useful for Nirmali. The administration failed to maintain and keep its investigation equipment functional. She had taken her husband to Chas Hospital in high hopes of getting appropriate treatment. He was already under the medication of Dr. Vimal Singh, but the household wanted to consult at a bigger hospital to get treated properly. She questioned why one should visit a government hospital when there is no facility. The lack of proper treatment facilities results in low health care utilisation given a doctor at the Chas hospital. It also affects the future annual funding of health services, as informed by the medical officer (MO) of Chas Hospital. "Medical officers keep requesting the civil surgeon to provide sufficient drugs and improve equipment and other facilities. There is no positive response from the higher authorities." One of the doctors at Chas Hospital said that they have to bear the brunt and face the patients every day. The patients who do not find the experience satisfactory do not return for follow up. A doctor at the government hospital follows a standard way of prescribing medicines and modulates the doses when it is required. This procedure requires more than one visit to the hospital, which causes hardship to the Hadis as each visit means a loss of a day's wage and transport costs.

Hadis also reported that doctors misunderstand the personal illness history of Hadis and the kinds of illnesses suffered by them. They do not give sufficient time and don't allow patients to describe the illness stories. One respondent, Ashuni Devi, aged 30 (wife of Manish Hadi), narrated, "Doctors continuously tried to console me *by saying that I do not have any problem. I felt like he was hurting and disrespecting me. I had pain in my stomach for a few weeks that took me to Chas hospital. Why would I otherwise visit a hospital so far from my house? It is difficult for me to trust those doctors after this incident. They do not have any feeling to hearing a patient in pain in a hospital, that too after waiting for 2 h outside his chamber. I lost a day's wages and could not prepare food for the young children at home. I asked the doctor where I was going from here. The doctor just shrugged his shoulders and said there was nothing he could do really." Ashuni Devi had gone to Chas Hospital to seek health care services for her stomach-related illness. She was not finding relief and cure even after taking Dr Vimal Singh's medication. Her experiences at the government hospital remind her of a bad day. She does not want to revisit Chas Hospital in the near future.*

The lack of effective communication with the doctor and Hadis is important to note. Effective communication between a patient and the provider is indeed supreme in a health care setting (Marrone 2007). It becomes rather complex in a society having caste-based social stratification and traditional occupational groups are functional.

The relationship between a Hadi patient and the caregiver is observed unequal at different sets of health care centres. The interaction is of an inferior and the provider as superior in the hospital setting. The supremacy over the knowledge and skills of biomedicine also influences the unequal relationship between a doctor and a patient. A patient finds little space and feels s/he gets less time to explain and narrate an illness felt, whereas the care provider, a doctor, does not give enough time to listening to the patient. Hadis speak a different language from the doctor. This concern is widely discussed in previous chapters as well. Even symptoms are described in medical terminologies alien to an uneducated patient like the Hadi. In this scenario, the Hadi patients have unpleasant and sometimes unsatisfactory experiences interacting with doctors, outside the Hadi Cooli vicinity, in any hospitals and health centres.

One does not find many Hadis who feel good after meeting the doctors. Ramesh Hadi recalls, "I was able to follow the language of doctors in the ward where I worked at BGH. The in-patient would mostly request me to explain the details of the illness in their language, as most doctors were unfamiliar with Khortha. A few nurses behave nicely to the poor patients. A patient keeps asking about every issue that troubles them as the in-patient already gives them a difficult time." Communication barriers deter Hadis visiting a hospital that appears alien and poses uncomfortable situations to them. In the case of Somnath Hadi, who had a fever and loose motion in the year 2010, due to his illness, he could not attend the day's work. I suggested he take two-three glasses of water with a spoon of sugar and salt. Fortunately, I, too, had two tablets of Combiflam (composition of brufen and paracetamol), mostly prescribed for fever and body ache. I gave him those tablets that he accepted humbly. Somnath swallowed one with a glass of water from his house as we were at the door. This provided some relief to him, and he could save at least 50 rupees for medicine (to express that he was obliged), as he told me, if he had visited the doctor. One has to pay for the consultation fees and *goli*, if given. The RMP (as doctor) are the best saviour for them. The minimal display of power and subordination between doctors and patients suits Hadis, unlike the clinics and hospitals outside Hadi Cooli.

Somnath recalled an unpleasant experience of visiting the government hospital a year back. He felt that doctors did not treat him properly. Somnath said, "When I visited Chas hospital in 2008 in the illness of my youngest son Sameer, aged six. Sameer had been vomiting and ran a high temperature. Dr. Vimal Singh was out of town for a week. I took a shared auto for 30 rupees for two passengers. Upon reaching the hospital, I made my son sit with his mother. I got a consultation slip from the counter. After an hour, we got a chance to meet the doctor. When we entered the consultation room, my son started vomiting. The doctor became furious. He quickly prescribed medicines without listening to the details of my son's illness. He asked his compounder to take us out. Though I cleaned the table with a towel, I had taken to cover Sameer. I was handed over a prescription with so many medicines. I had requested the doctor to give at least one bottle saline to my weakened son. Even the compounder did not help in this matter. He told me to buy medicines and return home." Somnath felt helpless and dejected for a while. He had high hopes to get better treatment. Dr. Vimal Singh never behaves like this. Without listening in detail, the doctor prescribed medicines. It is not perceived as proper treatment. He also hoped Sameer may be admitted due to his weakness and get saline water. He was briefed to take drugs if available at the drug store and return home. The medicine store was closed at that time, but he met a store person who said he had to buy the prescribed medicines outside. It was not available in store.

Meanwhile, his elder son, Laxman, who was also with them, went back to talk to the compounder. Laxman pleaded with him to talk to the doctor and provide better medicine or admit his brother in the hospital. At least one saline will help him to recover. It did not work. Lastly, he explained that medicines should be bought and returned home. Somnath was worried about his son. His condition was critical in the family's view. His wife suggested visiting KM Memorial Hospital, next to Hadi Cooli. She also expressed her willingness to sell her gold earrings, if required. Somnath returned home in an auto with his wife, Sameer and Laxman. He went to Chas's medical store to inquire about the cost of medicines. It was more than 500/– rupees. He bought medicines for 2 days. To his luck, the next day, Dr. Vimal Singh was available in his clinic. Upon examining the symptoms of illness and repeated vomiting, he advised one unit of saline to be administered immediately. He revised the prescription and asked Somnath to start the prescribed course of tablets. In an hour, Sameer started feeling better when saline was given. The vomiting sensation was reduced, and he had one glass of ORS powder orally after the saline was removed from the clinic.

The patient's family members passed a painful and helpless period during the entire episode of consultation at Chas Hospital until consulting Doctor Vimal Singh. He and his son Laxman lost one day's wage in addition to the cost of commutation. But their memories of humiliation and carelessness shown by the hospital doctor and compounder were alive. Afterwards, nobody from his house visited Chas hospital. Afterwards, he prefers consulting Doctor Vimal Singh. He does not ignore the narration of illness symptoms. This is most important to a patient, told Somnath. Health care provisioning is important, but it is not sufficient for health care utilisation. Various associated factors such as income status, caste, occupational association, gender, and locality of residence play a crucial role in influencing the perceived medical care utilisation (Gangadharan 2005).

It is commonly perceived that the distance to PHC and Chas Hospital is not a walking distance. Doctors are not available in the hospital. This is evident in a group discussion held in 2012 in Hadi Cooli, where people chose private hospitals located near their locality when they needed emergency medical care. They think there is assurance of treatment found that medical officers find under-staffed and no initiative for recruitment is taken. These two factors, understaffing and non-recruitment, would rectify medical care. The workload of the existing doctors and staff is higher than normal duty. The justification for the lack of services rests upon the above two factors that result and contribute to the loss of time and attention to patients, especially the poor and less educated ones, like Hadis. An inadequate number of doctors and support staff is justified for the low level of utilisation by Hadis. Therefore, the episodes of seeking health care at government hospitals demand higher expenditures of money, time, and a day's wage, which is saved in case one consults RMPs in the vicinity. Poverty indeed contributes to the inability of Hadis to pay for transportation and reach government hospitals. The cost of travel and transportation for those patients who were unable to walk increased. In Matthews (1979) study on health and culture in a south Indian village, she tries to understand the health culture of the various social groups in a village habited by both upper and lower castes. The Tamil habituated village was found to have unequal social distribution and differences in health culture in the context of access and utilisation of public health care services. Matthews found that 21% of village families were spending between 90% and 100% of their total income on food, and another 25% between 80 and 90% on health expenditure. Similarly, the expenditure patterns on food and health care by the poor Hadi households are high. Hadis report that there is no monthly calculation on the expenditure on health care because almost entire month's earnings are spent on providing food to the households. The remaining amount is spent on health care, marriage, and other kinship functions, religious rites, clothes, etc.

The additional factors of government hospitals, such as poor infrastructural set-ups, minimise their utilisation. The study of Banerji (1973) in his paper on 'Health Behaviour in Rural Populations: Impact of Rural Health Services' shows a different image of PHC, which is created based on lack of medicine, high crowding, long queues of people standing for their turns, and the socio-cultural gap between the patients and health services, a barrier in accessing them. The Hadis has quite often reported not visiting public health facilities due to above-mentioned reasons even though they are an urban population, unlike those in the study conducted by Banerji. On the other hand, one additional factor adds to the tyranny. Hadi individuals have better availability of both public and private health care services, but access to private health care in Chas town is easier than access to public health care services. Although the accessibility in private is easier than in public but at the cost of high expenditure, increasing the difficulty of the already burdened household. The socio-economic conditions of Hadis are such that the entire data on health and illness describe the inability to spend money on health care at private care centres and clinics.

Somesh Hadi states, "I worked for more than 20 years in BGH. My life was comfortable and in better working condition than my friends in Nagarpalika. Whenever someone visits Chas hospital, treatment is not provided. Another retired employee at BSL, Ramesh, recalls, "Handling medical waste, human waste and treatment-related material is to be thrown by the sweepers when I worked at BGH. The cleaning of medical waste harms and we may get injured. The company provided a uniform with a set of hand gloves. Often, sweepers do not prefer to use gloves. When we return, we change our uniform and dress our own clothes, then return home." The facilities of BSL are appreciated by retired employees of the steel plant. However, the number of employees at steel plant is limited to two persons during my fieldwork.

In contrast, when I discussed with the doctors in the hospital, they told me that the hospital does not get the required quantity of medicines from 'above' (above refers to the higher level of government functionaries). Medicines are not supplied regularly to the hospital. The district drug inspector cites a lack of resources for medical supplies as the main reason for the non-availability of free medicines. The non-availability of the district hospitals in Bokaro has increased the burden on small sub-divisional hospital hospitals in Bokaro has increased the burden on small sub-divisional hospitals. Public health care institutions are extremely overcrowded with patients. Also, compared to the number of patients accessing public health care institutions, the strength of the medical staff is low.

5.3 Experiences with Qualified Doctors and in Private Practice

The private hospitals in town are in good numbers. The hospital at Chas bus stand is closer than government hospitals. There are sufficient qualified doctors practising in Chas. But people do not go to these hospitals. Hadi patients are unable to pay the cost of care at private hospitals. It has increased in last few years. In many clinics and other private hospitals, consultation fees have increased from rupees 100 up to rupees 400 per visit except a few doctors who charge a fee of not more than Rs. 150 per visit. Due to high fees, Hadis are not able to utilise private health care services that turn them to RMPs or other healers. The experiences of failed medical care treatment are brought back to Hadi Cooli and construct perceptions of Hadis in accessing health care services.

The hereditary occupation of Hadis doubles the barrier. The data on the patient-doctor interaction between Hadis and hospital health personnel at private clinics and hospitals helps us understand the barrier. It was found that Hadis tried to bargain fees to be paid to doctors and other investigation costs. The fees seem very high to the Hadis. For instance, at the hospital called Rambha Hospital, also known as Srivastava's Clinic, a Hadi woman, Komna Devi, repeatedly requested to be exempted from the doctor's fee of Rs. 200. Her consultation with the doctor on the occasion of her daughter's's injured leg which happened while fetching drinking water one evening. During the discussion, Komna also argued that she cannot pay Rs. 800 for a small fracture. Consultation after the doctor's consultation, the remaining cost of Rs. 200 was for cement plaster (to bandage the fractured leg), injections and the cost of the material used. The hospital staff did not listen to her and asked her to meet the doctor himself for any discounts. The doctor was pretty unhappy with his staff allowing that lady into the hospital and warned them against repeating this in the future. The receptionist cum cashier, Guddu narrated the unhappiness of the doctor who also owns the hospital, "*Aage se konho Hadi-wadi ko Ilaje mat karo-nae to tum chod do naukri* (do not entertain any Hadi patient for treatment or you leave this job.)"

The doctor found it very offensive when the woman entered his chamber and made the doctor give her a discount by charging her half of the fees. The employee also shared that the doctor was unhappy and felt embarrassed when she did not listen to him. Rather, he pleaded for 10 min until Doctor Sahib (those in higher authority or position addressed with *sahib* as the suffix) agreed to give the concession.

In the above discussion, the fact that the woman was a Hadi emerged in the context of her pleading for concession, an identity which was hidden till then. After her conversation with the hospital staff and doctor, a feeling of discrimination emerges from both of them. Her behaviour is attributed to her caste and occupational background. This kind of conversation is often noted at clinics, private hospitals like Muskan and Neelam and PHCs. People make pleas at PHCs to get free medicines. Patients have a feeling that PHCs provide medicines at no cost to the poor. Such social interactions about the cost and fees are natural to other patients, but a barrier is created when it is associated with Hadis. I learnt from Guddu that sometimes doctor sahib gives concessions to poor patients if they plead nicely. He did not like her way of plea. She was continuously pressing him in his chamber despite being told that the fee could not be waived. He felt bad about the way Komna Devi persuaded.

A Hadi woman Pawani Devi, aged 30, says that she feels much deprived when she is unable to do her work, as it makes her dependent on other members of the family. She would never like to rest unless she has a serious illness. She narrates an incident when she delivered her first child with the help of a *Dai*. Unfortunately, she developed some complications and had to be admitted to the nearest private Neelam Nursing Home. During the treatment, she realised that it was the doctors' decisions only that would direct the patient's household members regarding the treatment and its care. The payment was to be done before the treatment started. She had to pawn her jewellery (which she received from her parents at her marriage) to a money lender for Rs. 15,000 to bear the hospital expenses. However, after 2 days, she was referred to another hospital in Ranchi. Non-compliance of Doctor Vimal Singh's advice left her upset and agonised. It took nearly a year to come out from the trauma of being hospitalised and the process of getting treatment, on top of all this, the baby did not survive. The apparent 25-year-old woman had not only lost her baby (the reason for which was not clear to her) but also was burdened with a huge rate of interest on the borrowed money due to hospital expenses. She is apprehensive about hospitals and modern, bio-medical care. She cannot think of getting admitted to any of the hospitals in the future, no matter what health urgency she faces. She still takes regular medicine according to the given prescription. Doctor Vimal Singh had advised Komda Hadi, Pawani Devi's husband, to inform him when delivery pain begins. He could oversee if necessary. The husband and wife discussed and decided to call Dai and did not inform the doctor in the clinic.

Her husband, Komda, works in a private office as a sweeper. He is an alcoholic, and generally, during the evenings, he spends time playing cards with his friends in Hadi Cooli. They have two children, a son and a daughter, who do not understand why their father drinks alcohol, cries and abuses their mother." This is not an unusual story to this household. Rather, it is a common fact among the Hadis. The Hadis in this region are dependent on alcohol and are habituated to gambling. This serious social problem has been noticed but barely any initiative has come from the state or civil society organisations.

Hadis are poor and are in a disadvantaged position. Their physical strength reduces with age, and reduced wages due to sickness have a deep impact on the economy of the entire household if someone like Nirmali's husband is ill for a longer period. Besides, they are not able to voice all their economic concerns. Before the doctors, these reasons become evident in Hadi's life when they have no money to approach a private practitioner (MBBS) or meet the cost of transporting a patient to a health institution in cases of emergency situations emergencies as against buying the prescribed medicine from the chemists' shop. Their desperate health needs are fulfilled in the form of *suya-goli* but it does not always serve the purpose in the long run.

The lack of paying capacity poor income, and drinking and gambling expenses dissuade them from accessing health care provided by the state, and the barriers to accessing the same are higher and thicker for them. The continuous association for generations to a menial occupation has not only affected the utilisation of health care services but also shows less social interaction with other castes. A vital consequence of this occupational association is the perpetual discrimination towards Hadis. Caste is an important consideration in health care access and utilisation, especially for occupational and scheduled caste groups like Hadis: socially marginalised and excluded group. Social and cultural marginalisation has shown a negative effect on health care. The traditional association with manual scavenging doubles the burden of social apathy towards Hadis as several Hadis felt health care was not received in a government hospital. Most Hadis turn towards private health care centres and clinics rather than public health centres.

The impediments to accessing health care services infringe the basic rights of Hadis to get health care support as a vulnerable section of society. Let's connect with the fact that "Vulnerability within the right to health framework means deprivation of certain individuals and groups whose rights have been violated from exercising agency" (Yamin 2005, p. 1).

Hadis are nowhere equal to all other castes. Their income, housing conditions, polluted occupation, social distance, perceived experiences of discrimination, etc., make them unequal to other castes. As discussed in Chap. 2, Hadis are also occupationally placed lowest among all the occupations in Indian society. A few past experiences of friends and relatives are mostly alive in the minds of households. It reappears in the decision-making process whether to avail government hospital services or not, and the last experiences, if any decision-making process whether to avail government hospital services or not, and the last experiences, if any, had. Many women do not visit government hospitals due to ill experiences either at the individual level or someone from Hadi Cooli. It is also perceived that at least consultation with the doctor is free, and sometimes certain medicines could be given, if available in a public health institution. Contrary to the popular perception of government hospitals, Hadis have not received free medicines at hospital.

There is an important role of education that is missing among Hadis. Lower educational level among them connects to a study's analysis in Kerala. It was found that higher women's education in Kerala has played a crucial role in improving health care utilisation due to a higher awareness level among them (Panikar 1979). In contrast, women in Hadi Cooli are less educated in comparison to the Hadi males. They are less informed, but women are quite familiar with Dr. Vimal Singh's clinic and find it most suitable for them.

Another study by Wan and Soifer (1974) brings up some important factors in health care services, such as illness level of the patients, average expenditure per visit, and age, apart from other indirect variables mentioned as deterrents. This study helps to see that the cost of per-visit consultation fees is important to any worker whose everyday income is crucial for the household. The cost of consultation and treatment not only refers to the fee paid to the doctor (in the case of a private qualified doctor), it also includes commutation or transport costs, and loss of wage. A study conducted at National Council for Applied Economic Research (NCAER) by R. Sunder (1995) finds that utilisation of health care services depended on distance of health centres with reasonably quality-based services and on the ability of the patients to avail those services.

In a discussion with Kabir Hadi and his spouse, Sarla, it was noted that even if the services are available, they need to be socially unbiased and free from prejudices. It is relevant to bring the story of Kabir and try to understand the impediments to accessing health care. Kabir and Sarla have not gone for the family planning method till now, although they have heard about family planning from ANM, Sima. A few couples at BGH adopt the family planning method with the help of ANMs-Sima, who sometime visits the Cooli or the government primary school.

Most front line health staff do not visit Hadi Cooli. Sometimes, an ANM visits the Cooli when it is important, and inform Ramesh Hadi. Ramesh's house is at the entrance to Hadi Cooli Street. It is convenient to visit his house. He is well-informed about the households in Hadi Cooli. Ramesh sometimes calls his caste members near his house, where health workers can inform and discuss the matter. Kabir said, "Check Post Hospital does give attention to family planning-related matters. There is one room where health workers provide information to go for sterilisation. We were also given counselling, but I did not visit again. I recall 4 years back when we went to consult the doctor for Shital, my elder daughter. They did not even provide any medicines to her. Recently, 1 month back, I again consulted for my headache and fever; a woman tried educating me that some amount would be given to me and my wife if we went for sterilisation, but I refused." There is a feeling that government hospitals do not address and resolve the problem, but the focus remains on family planning methods, not on treating patients.

The hospital staff does not even bother to about discuss the illness faced during the visit to health centres, unlike at the clinic of Dr. Vimal Singh. He is always sympathetic and empathises with the patients. Dr. Vimal Singh also charges consultation fee, and cost of medication, if given. But private hospitals charge high fees and would barely show empathy towards them. Recalling one of the experiences at a private health care Ganesh's wife narrated how she felt and the experiences in a private hospital. Sarla Devi said, "*Cheel near jhaap mare lago hathin yadi bujhe paral kau ke operation nae karle hiye. Operation ke upar badi zor de hathin kintu aare ilaaz nae miltau badhiya. Chas mein to private hospitalvo mein nursevan delivery time kahe lagthun ke kahil operation na karle hiin. Hamni garib parivar ke log daer ja hiye. Kakhno to jae ke man na karo hai par kate bach bin jael khatil kaam to milbe karo hai.* (The private hospital staff behave like vultures intensely searching for their food. Do not ask me about big private hospitals; they are never worried about the patients when I visit for some illness. They lay high emphasis on family planning operations but not on my illness. Less attention is given to the patients and their illnesses. The plight is similar in a private hospital when I visited for the second birth)."

Sarla said "Nurses began asking why she has not adopted family planning as if she has committed any crime. I got scared if they forced me to go for an operation." When she heard about the family planning operation, it was scary to her and challenging. She thought about why she visited the hospital. She could have avoided visiting hospitals if it was possible; but they get health troubles on and off. She has twins now. When the twins were to be born, Sarla was admitted to Neelam Nursing Home at Jodhadih More, Chas. The nursing staff and doctor also suggested adopting permanent sterilisation, but she was not ready. Most of the doctors told her that after having a twin baby why, she was not ready for permanent sterilisation, but how could she decide and give her decision? One has to think and decide properly about the problems and benefits. She gave up the idea to go for sterilisation. She also had a fear of losing her reproductive capacity. Sarla said at least one house can have three to four kids. She was scared of the tubectomy procedure. She was told by a friend that sterilisation weakens. Her husband was also not in favour. There is a perception that this procedure has health complications. A patient like Sarla Devi forms certain perceptions embedded in the socio-cultural processes and how people in the Hadi Cooli understand.

Sarla was not only scared and uncomfortable with the kind of social interaction between her and the hospital staff, nurses, and, at times doctors. But she also felt health providers do not empathise with the patients in the hospital setting. Hadi patients have less say over the health care providers, and how to interact with the hospital staff is different. In a few circumstances, when someone is unwell and falls sick, consultation is essential. In Sarla's case, they could afford the hospital expenses as her parents gave Rs. 20,000 for the childbirth in a nursing home. Financial help from her parents helped a lot to her.

The social segregation of the Hadis from other communities in terms of interaction impedes access to health care. To elaborate, the Hadis do not associate themselves with other castes residing in the neighbouring Mohalla. It is recorded in the data that they consciously refrain from engaging in any social relationship at the community level or at individual levels except on certain occasions as discussed in Chap. 2, in the youth and social interaction sections. A feeling of belonging to a manual scavenging caste dominates the physical movement of Hadis. Their occupation leads them to believe that they are born into a lower caste, and their occupation is not one of dignity. Another exciting fact recorded between Hadis and Bauris is that Hadis visit Bauri Cooli streets and attempt to interact for couple of reasons but not vice versa. First, Hadis get locally made liquor called *sasta daaru* (low-cost liquor). The sale and purchase of liquor happens in Bauri Cooli, bringing Hadis to the bylanes of Bauri Cooli. Second, a few youths go there to play cards with Bauris. But during field visits in the evening hours, Bauris were found to be playing cards with caste groups, which did not include Hadis. Ratu Bauri and Amulya Bauri, who were key informants among the Bauri Cooli during the fieldwork, argue that Bauri cannot go to Hadi Cooli as Hadis do dirty work that is impure and polluted. Upon questioning the similar status of both castes i.e. both of them fall under the SC category. They again take a stand not to talk about Hadis' life and stop further discussion. Ratu says, "*Atae gandh kaam ke paschaat barabari karon heen.*" ("After doing and attending such dirty jobs, how can there be equality").

The existing wide gap between Hadis and Bauris in social interaction is clear, though both are under the SC category. Bauris does not visit or attend any function or event in Hadi Cooli. In this context, Hadis are not given space for social interaction and communication with other castes. There is no denial that health care centres and clinics function in the same society. In this context, there impedes accessing health care furthered by a constant lack of interaction and unequal relationship between Hadis and others. Hadi patients also hesitate at the clinics of trained doctors, private hospitals, and government health centres or hospitals.

A few doctors were found to be saying in informal discussions that "Hadis will not change; they not only do filthy works but also stay unhygienic and unclean. Their behaviour is rude, and they don't know how to behave with a doctor." In contrast, many of the Hadis reported that doctors do not care to listen to our stories and illnesses. They do not want to give even 5 min to Hadis; what kind of expectations we will have? It is better we don't visit such government hospitals and avoid facing discrimination unnecessarily. Goli Hadi narrates, "*Hamni ke kaam kam general hai-ke-ab doctarvan ki gaari sunbin.*" Literally our work is so degraded that we again go to hear abuses from doctors? All these experiences bring Hadis the RMPs. The soft attitude of RMPs as doctors is a knack for consoling and managing patients, so they do return for follow-up consultation, which is used in maintaining a doctor-patient relationship. Most Hadis adopt a soft stand towards their local doctor in contrary to all other healers and trained doctors. The stereotyped image of Hadis as belonging to polluted occupation often emerges at hospitals and clinics outside the neighbourhood. Any wrongdoings are attributed to their caste and occupation.

Another dimension is about the desperation of the Hadis in seeking treatment. The flexibility of treatment fees is another important variable that popularises the existence of RMPs rather than visiting government hospitals or private physicians. The RMP does not fix his fees; rather the cost is decided according to the illness and desperation of the patients. On several occasions, the RMP doctors do not pressurise the patients to pay fees and they take it whenever patients can. Sometimes, doctors even note down the dues and collect them at the next meeting. The significance of desperation in getting cured of the illness provides a good scope to the doctor. They charge higher amounts in those situations. The patients need is only the patients only need a quick cure and recovery from the reported illness. They do not want to lose a day's wage due to illness.

In an encounter when Bajaj Hadi visited the clinic, his finger was bleeding from two cuts. He was injured while collecting garbage; he could not notice a naked blade which injured his fingers. Doctor Vimal Singh provided some liquid (Betadine), stopped the bleeding, and immediately applied a cotton and bandage. This treatment was followed by two *suya* (injections), one as a protection from tetanus and the second to reduce pain severity. After getting almost an hour of treatment time, Bajaj went back. I had long hours of discussion on this case when a few critical details emerged about seeking health-seeking perceptions and how an impediment emerges at other healing centres. Upon asking the Doctor Vimal Singh about the case, he said, "I did not even ask him for money. It is not right to ask and force him to pay when he was injured badly and bleeding. Second, he will be ready to pay a higher, or whatever I ask, amount if I charge because who will treat any patient on credit at this time? All Hadis know that private hospitals would not admit them and big doctors do not meet patients before taking consultation fees. It is only us, doctors in the neighbourhood, who give treatment on a credit basis. Even so I behave nicely." This shows the reason to prefer visiting a clinic in the locality over a bigger hospital. Immediate treatment is possible at both centres, but the charge fees of treatment are to be paid immediately before or after consultation at a bigger hospital. Vimal Singh's clinic would be paid consultation fees but at patients' convenience.

The problem lies more in the apathy of the concerned government officials, who do not represent their problem on most occasions. The medicine kits are given to ASHUNIs now as part of new initiatives of the Government of India known as the Urban National Health Mission (UNHM). I was informed by a few Bauris that ANM does visit the area, but she does not take rounds of household work regularly. It is their practice to collect information on women and the new born from key informants in the locality.

Lamti Devi, age, reported, "ANM's visit to Hadi Cooli is rare. She goes to the school and sits with the teachers. I remember sometimes she would pass by Hadi Cooli but would not make visits to the houses. They (health workers) come to my house as it is the first house when you walk into the street inside Hadi Cooli. Most government officials would ask my husband for some information. Who is free here? Most Hadis are busy with their daily livelihood." ANMs and *ASHUNIs* could have played a significant role in providing proper information related to government health initiatives and basic information. They are not regular to Hadi Cooli. Most Hadis reported not having seen both the health workers visiting Hadi Cooli. Hadis is not aware that ASHUNI health workers provide some medicines. Yes, a few women recalled having gotten *goli* from her.

While asking a few Hadis about the visits of ASHUNI workers, they could not understand who ASHUNI is. Hadis recalled ANM occasionally visited to their houses. She inquires about household details and talk to women at home. The absence of awareness about ANMs and health worker is noted. The constant presence of RMPs influences the health care seeking behaviour of Hadis. The increased gap due to their variance in cultural beliefs, behaviours, and values, patient perceptions of health and illness and health care practices may seem unfamiliar to the health care professional who practices outside the Hadi Cooli's surroundings. In other words, the convenience of visiting Dr. Vimal Singh's clinic is due to their familiarity and proximity with the doctor, less costly, time-saving, et cetera the doctor practising in the locality.

Hence, it is important to have better financial capacity to access even a private hospital and someone who can assist in accessing the hospital in case of any sudden health exigencies. For instance, a common Hadi is not well-informed about the probable expenses in treatment or pregnancy care until they are admitted and undergo the whole experience.

Ranjit is not different from most of his friends and peers in the community. The unseen facts that Sarla had revealed were generally ignored among the community and masses. Many of the family members perceive that the sense of responsibility is quite low among the government doctors and they do not care for the patients. In contrast, private consultants in private qualified doctors do charge handsome amount once a patient is admitted to a hospital. In a few occasions, they did show some courtesy to Hadi patients. One could make this difference from her narrations when she was not positive towards the behaviour and attitude shown by the government doctors. Contrary to her experience, the doctors at private nursing homes are more vocal and active while admitting the patients but in reality what matters is how much they can extract from the patients. If one cannot pay the whole amount, there is hardly any response to the patients and no care is ever possible. She admits that (*Sarkaari*) government-employed doctors do show some arrogance and rude attitude, but they seem more experienced and convincing. Above all, what matters is how treatment is done and whether there is any effort to the suffering.

5.4 Experiences with Traditional Healers

Further, the belief in traditional healers' practices (somewhat like magic, called, *tona-totka*[1]) is seen as helpful in their everyday lives. A few senior members in the community recalled a few forefathers wielding such magic. At present, it is found in villages. Hadis do not practice *tona-totka* themselves, but a couple of Hadis seek healing from such persons (shamans) who are available in the villages of Chandankiyari and its sub-urban area.

In one of the narratives, the role of tona-totka to heal was recounted. Somnath Hadi and Koyal Devi were graced with a son. In a discussion outside their house, they narrated the story of their second child. As told by Somnath, "Almost 5 years passed and we were still hoping for another child. At the suggestion of my parents, we visited a healer in the village, Mohanpur. We met an Ojha and narrated the purpose of the visit. Ojha usually performs rituals and provides herbal medicine prepared by him. He does not prescribe or disclose the name of the medicines to anyone. Thereby, he gave us doses of herbal medicine for 3 weeks. Upon the prescription to come again a month later, we revisited him. He performed rituals to provide solace to us. As asked, I bought one-metre of red coloured thread, five eggs, and one hen when we revisited. I do not know much detail as he sacrificed hen in the other corner of his backyard. Once he finished rituals, he returned a *handiya* (earthen pot) filled with ashes. Some broken eggs were kept in the pot mixed in ashes and tightly knotted in a small red cloth. He had asked us to sprinkle powder outside our house at midnight. We followed the suggested rituals religiously."

In the discussion, the influence of some *tona-totka* emerges. Those have command over *tona-totka* exercise over the others. Ojha had a feeling that Koyal Devi had bad

[1] A divine power bestowed on healers to treat certain illnesses is known as *Tona-Totka*.

influence of some magic (*tona-totka*[2]). Somnath described with great happiness about the positive outcome of Ojha's *tona-totka*. The ritual performance was perceived as having worked. The spouse was blessed with a son in a year. They also prayed, and their belief also depended on the blessings of Mansa Devi. There is no *suya-goli* to a woman not conceiving any issue. It is perceived as a problem of the spouse and household. The household discusses the reasons and deliberates upon seeking the help of Ojha or Baba.

I recall on my first visit to the Hadi Cooli area; they were all inside their houses; perhaps it was the wrong day to start off since everybody was inside their houses to avoid the blistering heat of the sun. The intense hot weather forced them to abandon their mid-day movements and activities. I found a person passing through the street half naked. He was wearing a half-folded *dhoti* (a white cotton cloth wrapped around the legs until the knees). The dhoti was white but was dirty and crumpled; it seemed that he had not washed it for a considerable time. Long-bearded, curly, and long-haired, he was walking barefoot. Soon it seemed he was given respect among the Hadis and other scheduled castes. Within minutes, I found that a small girl child was being blessed by this old man who appeared to be a saint or Baba (those who claim to be closer to God and have absolute dedication towards the almighty). While I was observing this person, another little kid immediately came running towards him. He would do nothing except touch his stick onto the head of the grace seekers. Local people believe that there is some grace that prevents them from illness if the Baba blesses them.

The existence and acceptance of such a person among Hadis is interesting. People do not bother much about his movements; rather, it was found to be a regular activity. I could see him on and off during the entire fieldwork period. It is known that he lived in a small village in Chandankiyari Block under Chas Sub Division Office. A few Hadis knew him for a couple of years when he would come to houses and collect whatever was given in the form of uncooked food and cash. During a group discussion, it was noted that the saint is a blessed man and gives his blessing to Hadis in crisis. But no significant charisma and divine action are seen or noticed in the life of Hadis. In other words, there is belief in his blessing, but Hadi youth are less impressed by Baba's charisma. In a group discussion, it was told, "Our parents respect him, so we do. We do not know his contributions and support in helping us. What we care about is that he is a Baba and respected in Hadi Cooli."

A traditional healer visits these areas from the nearby villages. They use some herbs that are unknown to the patients. Traditional healing patterns are accepted and availed by the healers who practice in the region. Abhilash and his wife have not been well for more than 2 years. They visited the traditional healers in areas of Chandankiyari, Phusro, and Purulia (West Bengal) also, but nothing helped Abhilash recover from the illness. There are many healers and Ayurveda doctors but all their efforts were in vain and ineffective. He wanted to get an immediate cure for the disease, a disorder in the testicles. The basic logic followed by Hadis is that illnesses

[2] *Tona-totka* is a kind of ritual performed to purify a human body from magic to restore normal health.

need to be cured quickly, and they need to get back to their daily activities. Also, realising that being ill or feeling unwell depends on the damage it has on their daily routine work.

The spouse, Abhilash Hadi and Sonal Devi visited Dr. Vimal Singh's clinic but it did not cure her illness. Just after returning from fieldwork, Abhilash Hadi took his wife to a *tantric.* He tried to heal her with all his tantric mantras. But these rituals, too, brought no relief. Once, he narrated, "If anybody goes to meet my wife, she starts crying and turns emotional. Whenever *Bhoot (an* evil spirit) possesses her body, she starts acting weirdly. She runs out of home and tries to jump into the dried well on the outskirts of Hadi Cooli. She also beats her children."

He further said, "*Ki kahiyo noonoo, hamar kanayyian naretiya dabae de ho* (at times, she tries to seize the necks of the children; perhaps she is depressed after her illness became serious. I consulted with a doctor at Ranchi, Bokaro. We spent a huge amount on her treatment and medication till now. A few kin members suggested they visit the Chiruaan dargah (Sufi *shrine)* in Giridih district. Every year in the month of Shabaan[3] the *Urs*[4] (annual programme) is held. Abhilash felt after staying 4 days at *the dargah, his wife's health had* improved. She tried harming herself on several occasions. When he returned from Dargah, on the advice of community members, he sacrificed a goat at *kaali mandir* and a hen.

Hadi's visit to Dr. Vimal Singh in seeking treatment at the clinics does not bring barriers, unlike in Chas Hospital or others. The pattern of availing health care services varies among Hadis. Pari Devi recalls, "I went to Telidih village and got some medicine for my prolonged pain and discomfort. I went under treatment by Dr. Vimal Singh to cure neck aches and joint pains. It never got cured permanently. The tablet subsided pain for a week, and then it would recur again. A friend of mine informed me about a person who provides treatment for these kinds of pains in the body. I consulted him (Jamuna Ji[5]). He provided some medicines that have reduced her problem a lot." It is also perceived among a few Hadis that the traditional healing system is less costly and most effective. In most cases, those patients who fail to find a solution in allopathic treatment seek alternative healing. But Hadis have got accustomed to modern medicine for their regular illness. It was reported that *suya-goli* is the best way of curing severe illness rather than taking ayurvedic medicines. Ayurvedic treatment takes a longer duration to cure any illness. It is felt that, in comparison, allopathic medicine provides immediate relief from absolute pain in the body or some discomfort. They get relief to attend their jobs. Buying allopathic *goli* at the medical shop is an easier task in order to manage recurring illness like fever, pain, and body ache.

---

[3] Shabaan is a month in Islamic Calendar (based on moon sight).

[4] *Urs* is the annual celebration on the occasion death anniversary of a Sufi in saint in India. Mostly it is practised in Sufi orders like Chishtiya, Qaudriya.

[5] Jamuna Ji lives in Cheera Chas, located at a distance of 4 km from Hadi Cooli. He provides herbal paste in certain joint pains and a few other illnesses. As he is aged more than 60 years, he cannot walk in forest to get his required leaf and herbs. Once a regular healer, Jamuna Ji tells his inability to provide herbal medicines.

5.5 Experiences with RMP Clinics

There is a shuffle between the RMPs clinic and Davaee Dukaan. While finding convincing treatment, Hadis visited Dr. Vimal Singh's clinic. The treatment at the clinic is quickly provided. The doctors give medication in high doses. I observed and heard most patients asking in the clinic, "Just give me some medicine to cure my illness. You know I need an effective treatment." Such conversation brings Hadis to the clinic, where they plead with Dr. Vimal Singh. A higher dose of tablets works more effectively than to begin with a low quantity of doses.

Tuberculosis bronchitis and lung-related diseases were reported among a few Hadis as Dr. Vimal Singh also confirmed. He provided treatment at his best capacity. He also said that people with cough symptoms are understood as having a minor illness. It is not easily identifiable to differentiate between cough and TB until certain investigations are conducted. The delay mechanism in reporting illness is common if it does not affect Hadis's work. It turns life-threatening sometimes.

During a visit at the clinic, I interviewed Shimti Devi. She had cough for 2 weeks. Initially, she did not perceive it to consult the Doctor. While asking why did you not visit doctor for your persistent cough? Shimti Devi told, "*Ki dakhaebin nunoo. Dalae halae ek barin syropva dactar thin gael haliya pache jadva main. Ab nae bae holae. Vaise baes roho hiye kintu jadvaa dinae mein kan-kani se khonkiya baedh ja hae. Noon-paani peeoo hiye jadva main ar ki karbin. Ab jael dihin jae bhatae*...(what shall I do my dear? Last winter, I visited Doctor Vimal Singh to seek relief from my increased cough due to severe cold weather. Usually, I would not have much problem, but it increases every year with the arrival of winter. I prepare and drink lukewarm water with some salt. How much can I go on consulting the doctor, let us see whatever happens)."

There is a perception that the treatment sought at Vimal Singh is sufficient to address illnesses like a cough. Moreover, women ignore seeking treatment unless it hinders their household and everyday routine. As in the case of Shimti Devi, she did not perceive her cough as problematic until she started having restlessness. Perceptions are necessarily personal, subjective, and contextual. It is immature because its own premises are not ordinarily subjected to critical appraisal (Sabharwal 2004). How one perceives reality becomes the basis for his/her actions and, therefore, goes into constructing the reality that may be perceived and vice versa. Hence, perception is the constructed reality of the society in the context of its subjective definition. For instance, Hadis perceive their surroundings in a way which is based on their experiences of living in society. The congested houses, filthy drains, and foul-smelling lanes/streets shape their day-to-day living and thought processes. Caste-based discrimination does influence the perception of the community. The environment of extreme poverty, social stigma, economic exploitation, and discrimination against untouchables (Hadis) is a reality for most in Hadi Cooli. Cases of bronchitis are very high among the Hadis in Hadi Cooli. Nine Hadis reported having bronchitis/lung-associated problems. Five persons were reported to have tuberculosis (TB) symptoms with a severe cough. But only two had been diagnosed with

TB. Skin-related infections would be commonly found, which is primarily due to sewage-related menial tasks. Dr. Vimal Singh is the most popular RMP with Hadis. Dr. Ansari and a few others are the other doctors with clinics in the vicinity. I also discussed with an Ayurvedic doctor who has a clinic cum medical shop at the main road market (discussed in Chap. 2). He has also observed that skin and bronchitis related problems are prevalent in the Hadi Cooli.

One evening, I was talking to a tea vendor near *Hadi Cooli.* Evening is one such occasion when several youths and old men gather at a tea stall near a junction, a narrow road passing by Hadi Cooli merges the main road, to have a cup of tea. I asked the tea vendor, "How many of the customers visiting at your tea stall would be from Hadis?" He said that most of the customers belong to Hadis and *neech jaati ke* (lower castes) from Bauri Cooli and a few from Muslim *Mohalla.* It was very informative and exciting to visit and have a cup of tea in the evening. This allowed me to meet people from the community very informally. Several visits to the tea shop helped me to build a good rapport with some of the younger Hadi boys in the vicinity of Hadi Cooli. After a month, I was less of an outsider. I was not viewed with suspicion and doubt. It gave me the strength to move positively and carry on with my fieldwork firmly.

I visited Vimal Singh's clinic, which was crucial to my fieldwork. I expressed my regards to the group and told them that I would be visiting them on several occasions and walked down. Pintu Hadi accompanied me. He was very happy and delighted to share his views and discuss them with me. I inquired about his family, work, income, and related questions, and we discussed them. Pintu Hadi stayed nearby waiting for me. He is a sweeper at BSL. He is uneducated as his father never encouraged his children to attend school. As already discussed earlier, children from *the Dalit community in earlier times were not allowed to attend school. The very thought of educating their children appeared to be horrifying as the teachers belonged to upper-caste groups. Further*, Hadis were not in a condition to bear the educational expenses of the children. Pintu Hadi said, "*Bhagwan Unch Jaat walon ko Itna balwan kyon banata hai, keval neechi jaati ke logo par atyachaar karte hain.*" Literally, why has god made the upper castes powerful, so only to oppress the downtrodden castes? The stigma and notion of belonging to degraded and untouchable caste still play a crucial role in their life. How does one feel if they have experienced discrimination in everyday life? Getting abused any time by the upper castes or finding a corner for a space in a public gathering is a common phenomenon in their life. He never received good wishes from upper-caste members. The only support was from his father, who said that Pintu Hadi should not worry and that fate would change 1 day. Pintu Hadi has been an honest person and believes in God. His general discussion made me think about the social stigma and a deeply entrenched repressive feeling among the Hadis (Dalits). He accompanied me to Dr. Vimal Singh's clinic, which was a little away from Hadi Cooli, in Bauri Cooli.

Vimal Singh is from the adjoining district, Giridih. He migrated from Giridih to *Bokaro* almost three decades ago. After his parents died, nothing was left at home and he was forced to search for some source of livelihood. He struggled to work at different occupations. He did not hesitate to work in a hotel as well. Dr. Vimal

Singh is known to most households residing in the neighbourhood of Cooli. In one of the visit to the clinic, there was a patient sitting outside the clinic. He was not conversant in Hindi. I had to manage talking to him in Khortha. When I asked him *"Kee bhello"* in Khortha, meaning "what happened"? He was amused and showed some inclination to respond to my question. Keeping the fact of language in mind I had to be ready as a field worker to face such circumstances when language may become a communication barrier. My immediate reaction was to shift to Khortha as my communication language during interaction with the community members. The patient did not reveal his name and even I didn't force him because he had already told me his residential locality. He was a wage earner. He had fever and headache for last 2 days. The previous day after observing his condition, he was given two tablets and asked to report back after a day. But his fever did not go down and he wanted to meet Vimal Singh again. Pintu Hadi, who was along with me, was surprised "Why I am asking and talking to people from his community?" He was surprised not because I was asking questions and trying to interact with them, but because the kind of questions being put to them was useless for him. He told me that generally, people coming for any survey would ask about family, how many children we have, income level, kind of houses, caste, etc. All these do not have any significance to their life and activity. Neither does this exercise change their impoverished condition; rather at times these serve as reasons for outsiders to humiliate the Dalit and poor.

After the discussion with Pintu Hadi and the patient, it was felt that the Dalit*s* are quite sensitive about caste hierarchy practised in society. They were conscious of being deprived in socio-economic spheres of their life. The same patient narrates "*Sarkar hamni ke kahan ate guda hau ke kohoon bes dactortheen jayeb, garib neechal jati ke kahan himmat hatao.*" The poor and lower caste can't afford to visit a good doctor at hospitals.

5.6 Concluding Words

As I have discussed, the Hadis' experiences at Chas Hospital are not uniform across the entire Hadi Cooli; these vary from person to person and household to household. For instance, the experiences and perceptions of BSL employees differ from the other Hadis. Yet, it must be mentioned that there is a great deal of overall similarity in perceptions regarding ill being, weakness, and access to health care facilities and availing them. The mismatch in working hours of Hadi workers and OPD timings at Chas Hospital, frequent non-availability of doctors, and sometimes also of the paramedical staff, and insufficient and/or lack of drugs, time, and money involved in reaching Chas Hospital are together recognised as a high-cost proposition by them. They lose out on both wages and cash by way of the cost of transport. Often the prescription of drugs instead of free dispensing from the hospital also disappoints them. The frequent non-availability of free drugs at the PHC is irksome. The advice for investigations means an additional burden on the household budget because often equipment at public hospitals is found to be out of order. Returning home without medicine and only prescription slips contradicts what Hadis expects from public hospitals. A large proportion of the Jharkhand population seeks private health care services, which are reported by NFHS III also, as mentioned earlier. For Hadis, the experience is more adverse as they feel being Hadis, they are discriminated against.

The experiences in utilising health care do shape their illness perceptions. Past experiences of illness drive the treatment and health-seeking behaviour. Despite having a PHC and one sub-divisional hospital in Chas town, Hadis find it difficult to procure satisfactory health care services when they occasionally manage to reach them. Additionally, given their experiences, they then are disinclined to go there again. Instead, they are dependent on (self-medication) at *davaee dukaan*, consulting RMPs and though it might mean paying out of pocket than at public health facilities for the treatment sought. As treatment sought at the medical shops and the clinics of RMPs are easily accessible in both social and physical ways, these constitute their familiar network and are easier to deal with, as discussed in Chaps. 3 and 4.

There are several qualified private practitioners in Chas town. However, private nursing homes and hospitals are not very approachable for Hadis for several reasons. The fees of private set-ups have increased over the last 10 years. Hadis could not think of visiting the nearest KM Memorial hospital at the Chas bus stand next to Hadi Cooli. But the consultation fee of a minimum of Rs. 400 per visit is beyond their reach. A couple of cases mentioned in this chapter elaborate on the burden the Hadis feel in seeking private health care.

Traditional healers also continue to have a presence in the life of Hadis in Hadi Cooli. Hadis approach traditional healers for specific illnesses. However, it needs to be pointed out that it is not less costly. They believe that some illnesses can be cured only by traditional healers. At times, the cost of treatment at the traditional healers' shrines or homes is substantial, but there is a belief in the efficacy of their healing that draws the Hadis towards them. Therefore, Suya-Goli is a metaphor, a reality, and an experiential part of everyday life for the Hadi people.

References

Adler, Nancy E. and Newman, Katherine. 2002. Socio-economic Disparities in Health: Pathways and Policies. *Health Affairs*. 21 (2): 60-76.

Banerji, D. 1973. Health Behaviour of Rural Populations: Impact of Rural Health Services. *Economic and Political Weekly*. 8 (51): 2261-2265.

Billings, John. *et al*. 1993. Impact of Socio-economic Status on Hospital Use in New York City. *Health Affairs (Milwood)*. 12 (1): 161-163.

Centre for Enquiry into Health and Allied Themes (CEHAT). 1995. Review of health care in India. CEHAT.

Gangadharan, K. 2005. *Utilisation of Health Services in Urban Kerala: A Socio-economic Study*. New Delhi: Serials Publications.

George, Sobin. 2015. *Caste and Care: Is Indian Healthcare Delivery System Favourable for Dalits*? Working Paper Series 350. Bangalore: Institute for Social and Economic Change.

Kulkarni, P. M., and Baraik, V. K. 2003. *Utilisation of health care services by Scheduled Castes in India* (Working Paper Series, No. 39). Indian Institute of Dalit Studies.

Matthews, D. A. 1979. Meaningful relationships: The role of friendship in medicine. *New England Journal of Medicine*, 300 (1), 39–41. https://doi.org/10.1056/NEJM197901043000109

Marrone, G. 2007. Factors influencing the use of sexual and reproductive health services among young men in Sweden. *BMC Public Health*, 7 (1), 195. https://doi.org/10.1186/1471-2458-7-195

Panikar, P.G.K. 1979. Resources not the Constraint on Health Improvement: A Case Study of Kerala. *Economic and Political Weekly*. 14 (44): 1803-1809.

Sabharwal, N. S. 2004. Caste, social exclusion and health deprivation in India. Indian Institute of Dalit Studies.

Sunder, Ramanmani. 1995. *Household Survey of Health Care Utilisation and Expenditure*. The Working Paper No. 53. New Delhi: National Council for Applied Economic Research.

Vijayanath, V. *et al.* 2010. Caste and Health in Indian Scenario. *Journal of Punjab Academy of Forensic Medicine and Toxicology*.(Online). 10 (1): 42-43. Acccesed on 18 May 2016, www.journals.pafmat.com/20101_11.php.

Wan, T.T.H. and S.J. Soifer. 1974. Determinants of Physician Utilisation: A Causal Analysis. *Journal of Health and Social Behaviour*. 15 (2): 100-108.

Yamin, Alicia Ely. (ed.) 2005. *Learning to Dance: Advancing Women's Reproductive Health and Well-Being from the Perspectives of Public Health and Human Rights*. p.1, cited from Chandirma Chatterjee and Gunjan Sheoran *Vulnerable* Groups In India Mumbai: CEHAT.

Chapter 6
Conclusion

In this dissertation, I have analysed the perceptions of illness and health among Dalits in Bokaro. Although the dissertation title uses Dalit as a generic category, my study is restricted to the Hadi Caste in Chas town. Hadis are a community of sweepers in the Municipal village of Chas, Bokaro. They are among the 22 SCs residing in Jharkhand (Census of India 2001). There has been relatively less attention in previous research on Hadis that could analyse the perceptions of illness and health. Further, the traditional engagement of this caste in manual scavenging distinguishes them from the other SCs in the region, which helped me to look at them ethnographically.

I have tried to provide a theoretical framework for illness, illness experiences, the health-seeking process, and understanding health as an integral component in people's lives. I also found an absence of ethnography on caste engaged in sweeping jobs (earlier manual scavengers), although quite a few studies Shyamlal (1984), Sachchidananda (2002), Chaplin (1997) and Ramaswamy (2005), and others have discussed manual scavenging and sweeper community and households. They provided data on caste and occupation, household structure, and social interaction with other castes, and the notion of pollution in their life. My focus relates to the sociological area of illness and health that has seen a great deal of attention in the recent past. The study has helped to understand the sufferings and experiences of illness and illness experiences that are brought effectively and objectively. The studies discussed in the introductory chapter have helped me understand health and illness better. My study has tried to understand the experiences and considerations in seeking health care.

Their experiences as Hadis and individual subjective position of Hadis as an individual influence their illness and health perceptions. It is a perception that depends upon the experiences of the illness and the illness reported in order to get relief.

The construction of illness perception begins when a Hadi worker feels that they are not feeling good and that they should rest for the day. Sometimes, it is also felt that it is viable to manage one day's absence from work and lose wages for that day.

K. M. Ziyauddin, *Occupational Caste Groups' Illness and Health in India*, People, Cultures and Societies: Exploring and Documenting Diversities,
https://doi.org/10.1007/978-981-96-8967-5_6

There is a constant dilemma for the Hadis. Various perceived feelings emerge among Hadi sweepers, and such narratives are discussed in Chaps. 3, 4, and 5.

I found Chas town in Bokaro to be the best possible site to conduct my fieldwork due to partial familiarity. The poor living conditions, social stigma, menial and traditional occupation, and segregation faced by the Hadis proved to be a challenge and added qualitative aspects to the contextualisation of illness. The Hadi Cooli is a largely marginalised urban locality, avoided by most social groups cutting across religious lines. These circumstances of the Hadis strengthened my resolve to take up this social group for intensive ethnographic study and come to a deeper understanding of the Hadi lifeworld.

My research examined the socio-economic status of the Hadis and used that to understand illness, illness perceptions and utilisation of health care services among Hadis. I have discussed how the socio-demographic profile and the occupational status of Hadis played a pivotal role in shaping their perceptions in the light of traditional occupation. It is imperative to remember that they are often housed in very congested houses on small plots of land, consisting of small rooms without any ventilation. My discussion of the basic amenities in the Hadi Cooli is reflective of the increased illness among Hadis. I have elaborated on the observation of doctors that Hadis live in congested houses and filthy streets and practice unhygienic practices such as open defecation, making them prone to higher chances of illness like viral infections and diarrheal problems. This further increases the morbidity among Hadis, who already tend to delay consulting a doctor and seek treatment. Such living conditions have no doubt contributed to the conditions of ill health and high morbidity among the Hadis. Therefore, the socio-economic conditions of Hadis appear to be an additional problem to their health and illness. However, further research is needed to understand more clearly the contribution of each of these factors and how they influence health care access and utilisation.

My central theoretical framework is on perception: Any perception is built in the context of Hadis' experiences in life. Hadi's perception is also an outcome of their occupational background as a sweeper. The working population among Hadis prefers to seek treatment for their perceived illness by taking *goli* and or *suya* by visiting Dr. Vimal Singh's clinic. Because seeking their treatment does prevent them from being away from their work. Consequently, Hadis thinks that the preference for treatment from these sources enables them to earn a daily wage and recover from an illness. Their household economy does not permit them to take on sick roles. The general pattern is to buy a *goli* to avoid remaining absent from work or to take a goli to avoid remaining absent from work. In a social context where Hadis are aware of the health care facilities in the town, they chose Davaee Dukaan and Doctor Vimal Singh's clinic near the locality.

The outcome of the study indicates that the experiences of an ill person are shaped by the sense of financial and social burden imposed by the illness; for instance, this might stop him from seeking treatment outside of bigger hospitals. Seeking treatment is not an individual's decision alone in most cases. When Hadi is the sole breadwinner in the household, the other members, especially women, suggest consulting a doctor and not just a temporary relief measure by buying *goli* over the counter. When women fall sick, they are more likely to find relief from OTC (over-the-counter) medicines rather than consulting by visiting the doctor at a clinic. This is not to suggest that men do not visit the *davaee dukaan*. Still, it is more likely that they will seek proper medical attention, i.e. by consulting the RMP at the nearest clinic, which is popular among all the Hadis in Hadi Cooli.

Hadis associate such feelings of being unwell and discomfort due to physical work they do with poor nutritional intake, smoking, and drinking. Nagarpalika employees struggle with illness and try to ensure that it does not deter them from work, and try to stay healthy to be fit to work. Even when they are very ill, they seek help from local practitioners rather than visit the more state-of-the-art hospitals, such as the ones available to employees of the Bokaro Steel Plant, in order to avoid missing work. The popular perception is that seeking help from qualified medical professionals often involves missing work, which is detrimental to earning their daily bread. My research also finds tremendous reliance on quick solutions such as injections and tablets, referred to as suya-goli, which they believe relieves pain quickly. The discussion on *suya-goli* emphasises a common connection between a health provider and the person seeking treatment. It also involves a higher level of Hadis's perceived satisfaction. In return, the doctor manages to earn a living through his services in the area. Quite often than not, *suya and related doses of suya are kept by a* doctor in a drawer in the clinic. The MRP (maximum retail price) of the medicine/injection is never shown to a patient, nor do patients ask to see them. There is a high level of trust between the doctor and the patients. The discussion during administering *suya* is interesting and would not take place elsewhere. In a situation when a patient is unaware and uninformed about the medicines, composition of the medicines and usefulness of medicine. The object (*suya*) and reference to *suya* heal a patient. Therefore, people in Hadi Cooli feel better once *suya* is administered. The role and function of Suya-Goli are socioculturally well-constructed in such a way that it helps patients with illnesses return to normalcy and is part of the social system, as Parsons mentioned in his theory of the sick role in 1951.

Women prefer consulting the doctor for their husbands or children rather than for themselves. Evaluation of the level of illness is essentially a significant aspect. Mechanic and Volkart's (1960) explanation of illness that an individual perceives certain symptoms and evaluates those perceived symptoms to act further on seeking treatment is relevant to the Hadis. Their evaluation of bodily symptoms helps Hadis decide whether to seek treatment. Based on the above considerations, a person would enter into the sick role (more often grudgingly) or attempt to retain normality.

My elaboration of three stages (Chap. 4) in seeking treatment reflects the intertwined interaction of the perceived illness by Hadis. Such a process is the theoretical outcome of this study that helped to reflect upon the perceived notion of illness and health among Hadis. I will try to explain further the illness experiences in the following sequence:

I have delineated the processes through which Hadi patients pass during the early stages of discomfort and illness and what determines the decision to seek or withhold treatment. Similar sequential illness experiences are significantly brought in Chap. 4 but are limited to perceiving health care services through *suya-goli*. Although the effort to bring illness experiences of Hadis analysed earlier in Chap. 3 helped to arrive at analysing illness perceptions. The perception of illness experiences is socio-culturally embedded among the Hadi households in Hadi Cooli. The occasional experiences of Hadis at hospitals and centres outside Hadi Cooli have also not been beneficial in seeking treatment. I argue that the experiences of Hadis go through a process that changes, influences, and perpetuates illness perception among them.

The data elaborates how the availability of hospitals and the costs of commuting to a location also deter them from seeking proper health care. The public health care services are in the form of PHC, at a distance of 3 km, and another sub-divisional hospital, at a distance of 5 km or even more in Chas, are the two referral centres to Hadis. Experiences in consulting these hospitals have not been taken as positively as they should have been. The loss of wages and time spent commuting to avail government hospital is not taken as a better means for cure. In addition, several past experiences have been found to deter people from accessing health care, both as individuals and as members of a low caste of sweepers with high expectations of professional diagnosis and medicines. Narratives of experiences are shared widely within the community and considered crucial, and they acquire pedantic power within the community.

Private hospitals are abundant in Chas (Bokaro) town compared to neighbouring districts like Dhanbad, Ramgarh, and Purulia in West Bengal. But Hadis do not have the spending power to go to such hospitals. Considering these circumstances, a strong feeling emerges that a doctor in the neighbourhood is the most appropriate centre for seeking treatment and cure. No comparison is drawn at the level of satisfaction and expenses incurred between a neighbouring doctor and a private hospital in Chas.

Moreover, the priority of visiting a doctor also has some other considerations. Important narratives have been brought in earlier chapters to describe a patient's feelings and how Hadis make decisions. Apart from consultation charges paid, some respect and dignity also emerge as one of the expectations among Hadis. Dignity, comfort, and self-respect have also influenced the minds' perception of Hadi's vis-a-vis. I have elaborated in the dissertation on how the occupations of Hadis have survived and lived notionally in their mind as unclean occupations. I also tried to examine the social history of Hadis, which kept them at a lower status at present than in their past when Hadis worked during the rule of the Raja of Kashipur.

My primary objective was not to elaborate on the accessibility and availability of health care services but to examine how Hadis constructed and perceived perceptions socially. The processes of the social construction of illness occur in the way illnesses are experienced and perceived: in the context of their daily activities, social interactions in and outside the Hadi Cooli, and the advice given to them at the medical shop and by the local RMPs, Dr. Vimal Singh and Dr. Pradip Kumar.

The entwined interaction between health care centres, *davaee dukaan*, Vimal Singh's clinic at Bauri Cooli, and health-seeking individuals has played an imperative role in influencing the behaviour of Hadi patients. Precisely in this context, access to health care services arises as an important dimension to understanding the illness behaviour and illness experienced by Hadis. In seeking health care, Hadis visits different care providers, and these are the sites where understandings of diseases are provided, shaped, and perceived, as discussed above. Given the background, my lay understanding of public health underlines the importance of how badly the poor need health care services that are easily available, accessible, and utilised properly. Banerji (1981a, b, c) argues that people's health patterns to utilise and avail of health care services are influenced by the political economy of the state: availability and location of the health facilities and social interaction with doctors and health personnel. Even when it is short, the mutual interaction between patients and the health care providers influences illness perception. Hadis and health care providers are mostly familiar with each other by face, caste, and locality of residence when consulting in the locality. I used two observation sites to develop my thesis—Davaee Dukaan and Doctor Vimal Singh's clinic during several short fieldwork stays from 2008 to 2013. I have also discussed the relationship between Hadis and care providers, in whatever brief form it has been for more than two decades. People are aware of the crude realities of life—Hadis-their association with menial and unclean jobs. They are unclean not only from the perspective of health and hygiene but also from caste pollution and ritual purity. As an ethnographer of the Hadi lifeworld, even I was subjected to this, as I recount in the Introduction chapter. This shows how such notions are deeply embedded in the community. For this reason, I was asked to vacate and was denied accommodation during my fieldwork in homes of other castes located close to Hadi Cooli.

Existing beliefs on goddess Mansa and Sitala Mata, values of community life, traditions, food practices, and perceptions of being well, behaviour, and religious ethos also shape how illnesses are perceived. I found that illness perceptions are not based on the medical understanding of illness but on perceived symptoms. The assessments of those symptoms are socially rooted in the community's life and work pattern.

Dr. Vimal Singh in Hadi Cooli often knows his patients' social and familial problems. The doctor's habit of probing his patients' personal and domestic issues is viewed as a positive aspect of their relationship with him. The inquiry of personal matters could have annoyed patients from the upper class who are aware and conscious of their social class. It is an endearing way of treating a patient in my study population. Such practices are taking the level of the relationship much deeper and continuing a process of rapport building. Thereby, the professional reliability of the doctor, even though they know he is not as qualified as other doctors found at private clinics and public health facilities in Chas: the mutual sociability between the two increases.

The experiences of illness are presented to see even small nuances in illness experiences and health-seeking at different stages. At a broader level, I have found it to be an eye-opening study on the Hadis. It could answer my questions raised in the initial period of my fieldwork about the first image of Hadi Cooli as filthy and unclean, the streets scattered with garbage, overflowing drains and clogged water, people playing cards on the street, and utensils washed on the narrow streets outside the houses Hadi boys hanging around the market area close by. People use filthy water to wash clothes at Joria, etc.

I have analysed the data on probable reasons for getting unwell, such as the absence of sanitation, the practice of open defecation, the lack of piped drinking water in homes in Hadi Cooli, etc. Hadis has not made a causal relationship to the problem of open defecation compared to observations of the health care providers in the locality. Experiences and incidences of increased viral infection and diarrheal cases possibly hint at those reasons. Encounters over the medical shop reflected the probable reasons for the higher reported illness in post-monsoon days. Problems of illnesses get answered at *davaee dukaan* through the kind of brief narrative presented. This form of interaction is equalitarian and much more influenced by the socio-cultural environment as they live nearby. The process of bringing symptoms and ill health is a perception, notions of illness, and probable ways to cure them.

Often, Hadis play an active role and create ways to avoid entering into the sick role unless it becomes a severe illness. The dissertation has argued that Hadis are mindful of the delay in reporting their illness. The primary concern to minimise the expenses on treatment is the significant outcome of the study. I have argued that an individual takes a long process before sitting idle because of an illness. A general perception among the workers does not allow them to accept the sick role unless it is beyond their control. Perceiving health care services through *suya-goli* concurs with the similar thought to delay reporting to the doctor. I find in the narratives of a few Hadis who report illness, seek health care and are diagnosed unwell, attract sympathy and support towards getting proper health care services to allow them to return to routine activity. The legitimisation of illness by the doctors is barely accepted among Hadis. The legitimacy after diagnosis provides certain benefits in the form of absence at work and less physical activity in the household. This framework is not profoundly reflected among Hadis in Hadi Cooli. The pressure to manage the needs of food, clothing, and health expenses of households keeps Hadis strong enough to continue engaged at the workplace. Illness is universal; one is familiar, but Hadis, in my research, would opt out of sick roles rather than taking rest instead of wages.

My ethnography on social stigma and discrimination shows that the level is decreasing compared to how it was described a few decades earlier. I have discussed in Chap. 3 the experiences Hadi patients bring back to their community and subsequent discussions flowing from them; it also shapes and influences the perceptions of Hadis. My elaboration reflects perceptions of Hadis' illness and health are constructed in their own experiences but not in isolation. The lack of empathy at government hospitals and the higher cost of treatment at private hospitals are additional barriers to accessing health care. Similarly, though without any consultation charges, their experiences of visiting government hospitals, PHCs, and SCs appeared to be a leading impediment that influenced Hadis's mindset about health care services. The cost of reaching distant public health facilities and the time spent seeking health care are also indirect. In other words, they face a loss of the day's wage. These calculations and monetary and time considerations pose a double loss to the Hadis, much to their dislike. These deter Hadis and turn them towards *davaee dukaan* and Dr. Vimal Singh's clinic, which is a stone's throw away. My deliberations in Chap. 5 have been to understand that even certain social facts impede accessing health care services in government hospitals and health centres. The experiences of discrimination felt by Hadis in the government hospital, the extra cost of transport to these hospitals, PHCs, and the several hours spent at these places seeking free medical treatment are essential social facts that discourage Hadis from seeking public health care. Similarly, a qualified private practitioner in town also deters Hadis as their consultation charges are considered very high by the Hadis.

6.1 Future Research

There is still scope that needs much deeper attention to study the illness perceptions of Hadis. I have attempted to bridge the shortage of study on Hadi's life and illness perception and broadly contribute to the sociology of health and illness. Much remains to be studied. Hadis has not been well investigated and would benefit from more social science research. Further research needs to be conducted to understand more about detecting occupational discrimination within the health care systems and provide validated data to the discipline of sociology on communities. One can also add to their study to explore and understand the health disparities of the socially most backward Hadis, considering their urban settlements in a Cooli pattern, as I mentioned earlier in Chap. 2. The social interaction with other SCs and the continuity of traditional occupation could be areas of future research. My study confines and focuses on studying the experiences of Hadis that help construct perceptions of illness and health. In a future attempt, the study on sanitation would further explore how Indian cities have developed. It has not been able to improve the sanitation facilities that have passed from the colonial period, similar to the argument put forth by Chaplin (1997) in her volume on 'The Politics of Sanitation in India: Cities, Services and the State'. She has attempted to study scavengers and how the sanitation problem in modern urban cities continues to demand manual scavengers even in the most advanced cities of India. I wish future studies would fill this gap.

References

Banerji, D. 1981a. Measurement of Poverty and Undernutrition. *Economic and Political Weekly.* 16 (39): 1579-1582.

Banerji, D. 1981b. The place of Indigenous and Western Systems of Medicine in the Health Services of India. *Social Science and Medicine.* 15A (2): 109-14.

Banerji, D. 1981c. *Poverty, Class and Health Culture in India.* New Delhi: Prachi Prakashan.

Census of India. 2001. Basic Data Sheet: District Bokaro- Jharkhand. (13), accessed 19 April, 2015. http://censusindia.gov.in/Dist_File/datasheet-2013.pdf.

Chaplin, Susan E. 1997. Scavengers: Still Marginalised. *The Administrator.* Vol. XLII (January-March): 123-151.

Mechanic, David and Edmund H. Volkart. 1960. Behaviour and Medical Diagnoses. *Journal of Health and Human Behaviour.* 1 (2): 86-94.
Ramaswamy, Geeta. 2005. *India Stinking: Manual Scavenging in Andhra Pradesh and their Work.* Chennai: Navayana.
Sachchidananda. 2002. *People at the Bottom: A Portrait of Scavengers.* New Delhi: Concept Publications.
Shyamlal. 1984. *The Bhangis in Transition.* New Delhi: Inter India Publications.

References

Acharya, Sanghmitra S. 2002. Health Care Utilisation in Rural North India-A Case of Nirpura District Meerut. An unpublished Project Report on '*Monitoring Shifts in Health Sector Policy in South Asia Project*'. New Delhi: CSMCH, JNU.

Acharya, Sanghmitra S. 2010a. Caste and Patterns of Discrimination in Rural Public Health Care Services. In Sukhdeo Thorat and Katherine S. Newman (eds.) *Blocked by Caste-Economic Discrimination and Social Exclusion in Modern India,* pp.208-29. Delhi: Oxford University Press.

Acharya, Sanghmitra S. 2010b. Access to Health Care and Patterns of Discrimination: A Study of Dalit Children in Selected Villages of Gujarat and Rajasthan. *Children, Social Exclusion and Development, Working Paper Series,* 1 (2). New Delhi: Indian Institute of Dalit Studies (IIDS).

Acharya, Sanghmitra S. 2011. Conceptualizing Social Discrimination in Access to Health Services towards Measurement: Illustrating Evidences from Villages in Western India. *Indian Emergency Journal*, VI (II): 14-27.

Acharya, Sanghmitra S. 2015. Access to Water and Sanitation Facilities-Missing Linkages with the Providers and Manifestations on Health. In S. B. Nimse and M. K. Agarwal (eds.) *Public Health and Development in India.* Pp. 26-45. New Delhi: Northern Book Centre.

Acharya, Sanghmitra S., Mala Mukherjee., and Kanhaiya Kumar. 2015. Social Inclusion in Universal Access to Antenatal Care: Understanding the Inter Group Inequalities. *Journal of Social Inclusion*, 2 (4), New Delhi: Indian Institute of Dalit Studies (IIDS).

Acharya, Sanghmitra S. and Golak B. Patra. 2016. Access to Maternal and Child Health Care Understanding Discrimination in Selected Slums of Delhi. In Sanghmitra S. Acharya et al. (eds.) *Globalising Delhi.* New York: Springer.

Adler, Nancy E. and Newman, Katherine. 2002. Socio-economic Disparities in Health: Pathways and Policies. *Health Affairs.* 21 (2): 60-76.

Aiach, P. and Curtis S. 1990. Social Inequalities in Self Reported Morbidity: Interpretation of Data from Britain and France. *Social Science and Medicine.* 31 (3): 267-74.

Anand, K., C. S. Pandav, and S. K. Kapoor. 2001. Injection Use in a village in North India. *National Medical Journal of India.* 14 (3): 143-4.

Anderson, R. and L. A. Aday. 1978. Access to Medical Care in the U.S. Realised and Potential. *Medical Care.* 16 (7): 533-46.

Anderson, Ronald M. 1995. Revisiting the Behavioural Model and Access to Medical Care: Does It Matter. *Journal of Health and Social Behaviour.* 36 (1): 1-10.

Apple, D. 1960. Sociological Studies of Health and Sickness: A Source Book for the Health Professions. McGraw- Hill. SAGE Journals.

Armstrong, D. 1984. The Patient View. *Social Science and Medicine.* 18 (9): 737-44.

K. M. Ziyauddin, *Occupational Caste Groups' Illness and Health in India*, People, Cultures and Societies: Exploring and Documenting Diversities,
https://doi.org/10.1007/978-981-96-8967-5

Atluri, Tara. 2011. You Marxist I Clean Toilet: Racism, Labor, and the Bathroom Attendant. *Frame: A Journal of Visual and Material Culture.* Spring (1): 69-95.

Ayers, Stephanie L., and Jennie Jacobs Kronenfeld. 2007. Chronic Illness and Health-Seeking Information on the Internet. *Health.* 11 (3): 327-347.

Banerji, D. 1973. Health Behaviour of Rural Populations: Impact of Rural Health Services. *Economic and Political Weekly.* 8 (51): 2261-2265.

Banerji, D. 1971. *Family Planning in India: A Critique and A Perspective.* New Delhi: People's Publishing House.

Banerji, D. 1974. Social and Cultural Foundations of Health Services Systems. *Economic and Political Weekly.* 9 (32/34): 1333-1346.

Banerji, D. 1978. Political Dimensions of Health and Health Services. *Economic and Political Weekly.* 13 (22): 924-928.

Banerji, D. 1981a. Measurement of Poverty and Undernutrition. *Economic and Political Weekly.* 16 (39): 1579-1582.

Banerji, D. 1981b. The place of Indigenous and Western Systems of Medicine in the Health Services of India. *Social Science and Medicine.* 15A (2): 109-14.

Banerji, D. 1981c. *Poverty, Class and Health Culture in India.* New Delhi: Prachi Prakashan.

Banerji, D. 1984. Primary Health Care: Selective or Comprehensive. *World Health Forum.* 5 (4): 312-5.

Banerji, D. 1989. Rural Social Transformation and Changes in Health Behaviour. *Economic and Political Weekly.* 24 (26): 1474-1479.

Baru, Rama. *et al.* 2010. Inequities in Access to Health Services in India: Caste, Class and Region. *Economic and Political Weekly.* 45 (38): 49-58.

Benyamini, Yale. 2011. Health and Illness Perceptions. In Howard S. Friedman (ed.) *The Oxford Handbook of Health Psychology.* pp. 285-318. New York: Oxford University Press.

Berman, P., R. Ahuja., and L. Bhandari. 2010. The Impoverishing Effect of Healthcare Payments in India: New Methodology and Findings. *Economic and Political Weekly.* 45 (16): 65–71.

Bhardwaj, Ruby. 2010. Medical Pluralism in India: The Interface of Complementary and Alternative Therapies with Allopathy. In Arima Mishra (ed.) *Health, Illness and Medicine: Ethnographic Readings.* New Delhi: Orient Blackswan.

Bhore Committee. 1946. A study of the health needs of the people and their implementation. Government of India.

Bhore, Joseph William. 1946. *Health Survey and Development* Committee (Bhore Committee). New Delhi: Government of India.

Billings, John. *et al.* 1993. Impact of Socio-economic Status on Hospital Use in New York City. *Health Affairs (Milwood).* 12 (1): 161-163.

Bir, Thaneswar. 2001. *Dynamics of Health Culture: Urban Slum and Behaviour.* New Delhi: Rajat Publications.

Blane, D. *et al.* 1996. Illness Behaviour and the Measurement of Class Differentials in Morbidity. *Journal of the Royal Statistical Society.* 159 (1): 77-92.

Bowling, A. 1991. Measuring health: A review of quality of life measurement scales. Open University Press.

Bury, Michael. 1982. Chronic Illness as Biological Disruption. *Sociology of Health and Illness.* 4 (2): 167-182.

Carstairs, G. M. 1955. A village in Rajasthan. In M. N. Srinivas (Ed.), *India's villages,* 36–52. Government of West Bengal Press.

Caudill, William. 1959. Culture and Personality: The Psychiatric Hospital as a Small Society. Reviewed by Ray Wilkie *American Anthropologist.* 61 (4): 714-716.

Caudill, William. 1956. Perspectives on Administration in Psychiatric Hospitals. *Administrative Science Quarterly.* 1 (2): 155-170.

Census of India. 2001. Basic Data Sheet: District Bokaro- Jharkhand. (13), accessed 19 April, 2015. http://censusindia.gov.in/Dist_File/datasheet-2013.pdf.

Census of India, 2011. *District Census Handbook.* Jharkhand Series-21. Part XII-B. New Delhi: Registrar General of India.

Centre for Enquiry into Health and Allied Themes (CEHAT). 1995. Review of health care in India. CEHAT.

Chaplin, Susan E. 1997. Scavengers: Still Marginalised. *The Administrator.* Vol. XLII (January-March): 123-151.

Charmaz, K. 1999. Stories of suffering: Subjective tales and research narratives. *Qualitative Health Research*, 9 (3), 362–382.

Chas Master Plan, Vision 2027. Chas Municipal Corporation. accessed 10 August 2015. http://chasmunicipalcorporation.com/AboutCMC.aspx.

Chas Municipal Corporation. 2027. GIS-based master plan for Chas. *Urban Development & Housing Department*, Government of Jharkhand.

Chatterjee, Chandrima and Gunjan Sheoran. 2007. The Public health-Human Rights Dialogue. In *AIDS, Health and Human Rights: An Explanatory Manual.* Section 4: 39. cited from Vulnerable Groups in India. Mumbai: CEHAT.

Chaudhuri, B. 1990. *Cultural and Environmental Dimensions of Health.* New Delhi: Inter-India Publications.

Coleman, Rebecca M. 2008. The Human Right of Sanitation for All: A Study of India. *Global Business and Development Law Journal.* 24 (1): 267-302.

DiMatteo, M. Robin and Freidman, Howard S. 1982. A Model Course in Social Psychology and Health. *Health Psychology.* 1 (2): 181-193.

Dimiuou, N. 1995. Illness and Culture: Learning Differences. *Patient Education Counsseling.* 26 (1-3): 153-7.

District Statistical Handbook, Bokaro. (2014). *Government of Jharkhand.* bokaro.nic.inindiastatdistricts.com

District Statistical Handbook, Purulia. (2014). *Department of Planning and Statistics*, Government of West Bengal. Wikipediapurulia.gov.in

Djurfeldt, Goran. and Stefan Lindberg. 1980. *Pills Against Poverty: A Study of Introduction of Western Medicine in a Tamil Village.* New Delhi: Macmillan India.

Early, E. A. 1982. The Logic of Well Being: Therapeutic Narratives in Cairo, Egypt. *Social Science and Medicine.* 16 (16): 491-97.

Elliot, J. 2005. *Using Narrative in Social Research: Qualitative and Quantitative Approaches.* London: Sage Publications.

Faircloth, C. A., Boylstein, C., Young, M. E., and Gubrium, J. 2004. Sudden illness and narrative flow in narratives of stroke recovery. *Sociology of Health & Illness*, 26 (2), 242–261.

Felski, Rita. 2000. The invention of Everyday life. *Cool Moves-A Journal of Culture Theory Practice.* 39 (9): 15-31.

Frank, Arthur W. 1994. Reclaiming an Orphan Genre: The First-person Narrative of Illness. *Literature and Medicine.* 13 (1): 1-21.

Frank, Arthur W. 1995. *The Wounded Storyteller: Body, Illness, and Ethics.* Chicago: University of Chicago Press.

Frank, Arthur W. 1997. Illness as Moral Occasion: Restoring Agency to Ill People. *Health: An Interdisciplinary Journal of Social Study of Health, Illness and Medicine.* 1 (2): 131-148.

Frank, Arthur W. 1998. Just listening: Narrative and Deep Illness. *Families, Systems and Health.* 16 (3): 197-212.

Freeman, James. 2004. *Collecting the Life Histories of an Indian Untouchable.* In Vinay Kumar Srivastava (ed.) Methodology and Fieldwork. New Delhi: Oxford University Press.

Funch, Flemming. 1995. *Transformational Dialogues: Facilitator Training Manual.* No. 1. California: Transformational Processing Institute.

Gangadharan, K. 2005. *Utilisation of Health Services in Urban Kerala: A Socio-economic Study.* New Delhi: Serials Publications.

Garg, C. C., and Karan, A. K. 2009. Reducing out-of-pocket expenditures to reduce poverty: A disaggregated analysis at rural–urban and state level in India. *Health Policy and Planning, 24*(2), 116–128. https://doi.org/10.1093/heapol/czn046

George, Sobin. 2015. *Caste and Care: Is Indian Healthcare Delivery System Favourable for Dalits?* Working Paper Series 350. Bangalore: Institute for Social and Economic Change.

Goffman, Erving. 1961. *Asylums: Essays on the Social Situations of Mental Patients and Other Inmates.* Chicago: Aldine Publishing Co.

Government of India (GoI). 2014. National health profile 2014. Central Bureau of Health Intelligence, Ministry of Health and Family Welfare, Government of India.

Gregory, Richard L. 1987. Perception. In R.L Gregory and O L Zangwill (eds.) *The Oxford Companion of the Mind*, pp. 598-601., Oxford: Oxford University Press.

Gregroy, L. Weiss and Lonnquist E. Lynne. 2012. *Sociology of Health, Healing, and Illness* (7th edition). Boston: Prentice Hall.

Grover, Shalini. 2007. *Looking Inside Marriage: Lived Experiences, Notions of Love and Kinship Support amongst Working Women in New Delhi.* Working Paper Series No. E/285. New Delhi: Institute of Economic Growth.

Hannay, David. 2011. The Symptom Iceberg. *British Journal of General Practice.* 61 (584): 224.

Haron, Yafa., Eisikovits Rivka., and Shai Linn. 2004. Traditional Beliefs Concerning Health and Illness among Members of the Circassian Community in Israel. *Journal of Religion and Health.* 43 (1): 59-72.

Harshe, Rajan. 2014. Review-A History of Prejudice: Race, Caste, and Difference in India and the United States, accessed on 7 Aug, 2016. www.cambridgeindia.org/reviews/Seminar

Husserl, Edmund. 1970. *Crisis of European Sciences and Transcendental Phenomenology.* Evanston: North Western University.

Hyden, G. C. 1997. Civil society, social capital, and development: Dissection of a complex discourse. *Studies in Comparative International Development, 32*(1), 3–30. SpringerLink. https://doi.org/10.1007/BF02696304

Ilaiah, K. 1998. Why I am not a Hindu: A Sudra critique of Hindutva philosophy, *culture and political economy.* Samya.

International Institute for Population Sciences (IIPS) and ORC Macro. 1995. National Family Health Survey (NFHS–1), 1992–93: India. *IIPS.*

International Institute for Population Sciences (IIPS) and ORC Macro. 2000. National Family Health Survey (NFHS–2), 1998–99: India. IIPS.

International Institute for Population Sciences (IIPS) and ORC Macro. 2007. National Family Health Survey (NFHS–3), 2005–06: India. IIPS.

Jaiswal A., N. K. Patro. and C. S. Pandav. 2006. Occupational Health and Safety: Role of Academic Institutions. *Indian Journal of Occupational Environment.* 10 (3): 97-101.

Jharkhand Municipal Act. 2011. Jharkhand Act 07 of 2012. Ranchi: *Jharkhand Gazzette.*

Jhosua Project. 2014. *Hadi in India*, accessed on 14 Jan, 2015. https://joshuaproject.net/people_groups/12027/IN.

John B. Mckinlay and Mckinlay Sonja. 1997, reprint 2005. Medical Measures and the Decline in Mortality. In Peter Conrad *The Sociology of Health and Illness: Critical Perspectives*, pp. 7. New York: Worth Publishers.

Julia E. *et al.* 1989. Health Perceptions of Primary Care Patients and the Influence on Health Care Utilization. *Medical Care.* 27 (3): 99-109.

Kanjila, Barun *et al.* 2007. *Health Brief: A Research Initiative*, June, Jaipur: IHMR. p.1.

Karlekar, M. (1976). Poverty and women's work in India. Concept Publishing Company.

Katavic, Snjezana *et al.* 2015. Illness Perception and Information behaviour of Patients with rare chronic diseases.

Keith, D. A., *et al.* 2021. Indicative distribution maps for ecological functional groups—Level 3 of IUCN global ecosystem typology, version 2.0.1b. Zenodo. https://doi.org/10.5281/zenodo.1234567

Kelly, M., and Field, D. 1996. Medical sociology, chronic illness and the body. *Sociology of Health & Illness*, 18 (2), 241–257.

Kulkarni, P. M., and Baraik, V. K. 2003. *Utilisation of health care services by Scheduled Castes in India* (Working Paper Series, No. 39). Indian Institute of Dalit Studies.

Kumar, V. (2005). Situating Dalits in Indian Sociology. Sociological Bulletin, 54(3), 514–532. http://www.jstor.org/stable/23620623

Rare Chronic Diseases. (n.d.) *Information Research-An International Electronic Journal*. 21 (1): 1-28.

Kermode, Michelle. and Vanlal Muani. 2006. Injection Practices in the Formal and Informal Healthcare Sectors in Rural North India. *Indian Journal Medical Research*. 124 (November): 513-520.

Khare, R. S. 1984. *The Untouchable Himself: Ideology and Identity*. Cambridge: Cambridge University Press.

Kleinman, Arthur. 1980. *Patients and Healers in the Context of Culture: An Exploration of the Borderland between Anthropology, Medicine, and Psychiatry*. California: University of California Press.

Kleinman, Arthur. 1985. Interpreting Illness Experience and Clinical Meanings: How I See Clinically Applied Anthropology. *Medical Anthropology Quarterly*. 16 (3): 69-71.

Kleinman, Arthur. 1988. *The Illness Narratives: Suffering, Healing, and the Human Condition*. New York: Basic Books.

Klinenberg, Eric. 1997, reprint 2005. Dying Alone: The Social Production of Urban Isolation. In Peter Conrad (ed.) *The Sociology of Health and Illness: Critical Perspectives*, pp. 159-170. New York: Worth Publishers.

Koos, Earl L. 1960. Illness in Regionville. In Dorrian Apple (ed.) *Sociological Studies in Health and Sickness*, Pp. 9-14. New York: McGraw Hill.

Kosa, J. and L. Robertson. 1969. The Social Aspects of Health and Illness. In Kosa et al. (eds.) *Poverty and Health: A Sociological Analysis*, Cambridge: Harvard University Press.

Kotwal A. *et al.* 2004. Injection Practices in a Metropolis of North India: Perceptions, Determinants and Issues of Safety. *Indian Journal of Medial Science*. 58 (8): 334-44.

Krauss, Bjorn. 2015. 'The Life We Live and the Life We Experience: Introducing the Epistemological Difference between 'Lifeworld' (Lebenswelt) and 'Life Conditions' (Lebenslage). Social Work and Society. 13 (2).

Kumar, Ashish and Dilip Kumar Sinha. 2014. *Opposite Behaviour of Food Market from that of Other Commodities: A Cause of High Inflation*. A Working Paper, accessed on 22 Oct, 2016. http://mospi.nic.in/mospi_new/upload/paper.

Kumar, Avinash and Meerambika Mahapatro. 2010. Narrator and the Narrative: Understanding Scientific Realism of Chronic Diseases in the Field. In Arima Mishra *Health, Illness and Medicine* (ed.) New Delhi: Orient Longman.

Labov, W., and Waletzky, J. 1967. Narrative analysis: Oral versions of personal experience. In J. Helm (Ed.), *Essays on the verbal and visual arts*, 12–44. University of Washington Press.

Lahiri-Dutt, Kuntala., Radhika Krishnan, and Nesar Ahmed. 2012. Land Acquisition and Dispossession Private Coal Companies in Jharkhand. *Economic and Political Weekly*. XLVLI (6): 39-45.

Last, J. M. 1963. The iceberg: Completing the clinical picture in general practice. *The Lancet*, 282 (7297), 28–31. https://doi.org/10.1016/S0140-6736(63)92717-6

Light, Donald and Alexander Schuller. 1986. The Political Values and Health Care: The German Experience. cited from William C Cockerham *Medical Sociology*, Englewood Cliffs: Prentice Hall.

Lok Sabha Secretariat Parliament Library and Reference, Research, Documentation and Information Service (Larrdis). 2013. *Manual Scavengers: Welfare and Rehabilitation-Reference Note for Members of Parliament*, No. (18): Rn/Ref./August.

Mahapatro, Meerambika and A. K. Kalla. 2000. Health Seeking Behaviour in a Tribal Setting. *Health and Population- Perspectives and Issues*. 23 (4): 160-169.

Marrone, G. 2007. Factors influencing the use of sexual and reproductive health services among young men in Sweden. *BMC Public Health*, 7 (1), 195. https://doi.org/10.1186/1471-2458-7-195

Mary, Christopher O'Rourke. 1960. Review: Sociological Studies of Health and Sickness. by Dorrian Apple, *The American Catholic Sociological Review*. 21 (2): 173.

Massey, James. 1997. *Down Trodden: The Struggle of India's Dalits for Identity, Solidarity and Liberation*. Geneva: WCC Publications.

Mathur, Indu and Sanjay Sharma. 1995. *Health Hazards, Gender, and Society*. Jaipur: Rawat Publications.

Matthews, D. A. 1979. Meaningful relationships: The role of friendship in medicine. *New England Journal of Medicine*, 300 (1), 39–41. https://doi.org/10.1056/NEJM197901043000109

McAteer, H., Elliott, J., Hannaford, P. C., and Heaney, D. 2011. Evaluating web-based self-management support for people with chronic pain: A randomized controlled trial. Pain, 152 (3), 833–842. https://doi.org/10.1016/j.pain.2010.12.008

McBroom, William H. 1970. Illness, Illness Behavior and Socio-economic Status. *Journal of Health and Social Behavior*. 11 (4): 319-326.

McKeown, Thomas. 1976. *The Modern Rise of Population*. New York: Academic Press.

Mckeown, Thomas. 1979. *The Rise of Medicine: Dream, Mirage and Nemesis*? London: Nuffield Provincial Hospitals Trust.

Mechanic, David. 1968. *Medical Sociology: A Selective View*. New York: Free Press.

Mechanic, David and Edmund H. Volkart. 1960. Behaviour and Medical Diagnoses. *Journal of Health and Human Behaviour*. 1 (2): 86-94.

Mechanic, David. 1986. The Concept of Illness Behaviour: Culture, Situation and Personal Predisposition. *Psychological Medicine*. 16 (1): 1-7.

Meleis, A. I. 1990. Theoretical nursing: Development and progress (2nd ed.). J. B. Lippincott.

Mendelson, Cindy. 2002. Health Perceptions of Mexican American Women. *Journal of Transcultural Nursing*. 13 (3): 210-7.

Michael, S M. 2007. *Dalits in Modern India: Vision and Values*. New Delhi: Sage Publications.

Ministry of Drinking Water and Sanitation. 2013. *A Document-Construction of Toilets*. New Delhi: Government of India.

Ministry of Health and Family Welfare (MoHFW). 2014. *National Health Policy 2015 Draft*. New Delhi: Government of India.

Minocha, Aneeta. 1980. Medical Pluralism and health Services in India. *Social Science and Medicine*. 14B (4): 217-23.

Mishler, Elliot G. 1986. *Research Interviewing: Context and Narratives*. Cambridge: Harvard University Press.

Mishra, Arima. 2010. *Health, Illness and Medicine: Ethnographic Readings*. New Delhi: Orient Blackswan.

Misra, Babagrahi. 1969. Sitala: The Small Pox Goddess of India. *Asian Folklore Studies*. 28 (2): 133-142.

Morrell, D. C. and C. J. Wale. 1976. Symptoms Perceived and Recorded by Patients. *Journal of the Royal College of General Practitioners*. 26 (167): 398-403.

Mumby, D. K. (Ed.). 1993. *Narrative and social control: Critical perspectives*. SAGE.

Murray, S. A., Kendall, M., Boyd, K., and Sheikh, A. 2005. Illness trajectories and palliative care. *BMJ*, 330 (7498), 1007–1011.

Nayak, Pulin. 1995. *Social Exclusion in India*. Geneva: International Institute of Labour Studies.

Nicholas, Ralph W. 1981. The Goddess Sitala and Epidemic Smallpox in Bengal. *Journal of Asian Studies*. 41 (1): 21-44.

Pandey, Gyanendra. 2013. *A History of Prejudice: Race, Caste, and Difference in India and the United States*. New Delhi: Cambridge University Press.

Panikar, P.G.K. 1979. Resources not the Constraint on Health Improvement: A Case Study of Kerala. *Economic and Political Weekly*. 14 (44): 1803-1809.

Pargament, Kenneth I. *et al.*2004. Religious Coping Methods as Predictors of Psychological, Physical and Spiritual Outcomes among Medically Ill Elderly Patients: A Two Year Longitudinal Study. *Journal of Health Psychology.* 40 (4): 126-134.

Patel, Tulsi. 2008. Stigma Goes Backstage: Reservation in Jobs and Education. *Sociological Bulletin.* 57 (1): 97-114.

Patel, Tulsi. 1994, 2006. *Fertility Behaviour: Population and Society in a Rajasthan Village.* New Delhi: Oxford University Press.

Pathak, Bindeshwar. 1991. *Road to Freedom: A Sociological Study on the Abolition of Scavenging in India.* Delhi: Motilal Banarsidas Publishers Pvt. Ltd.

Patro, P. K., and Acharya, S. K. 2016. Utilization of ferrochrome wastes such as ferrochrome ash and ferrochrome slag in concrete manufacturing. *Waste Management, 58,* 73–81. https://doi.org/10.1177/0734242X16654751

Pelser, Adam. 2010. Belief in Reid's Theory of Perception. *History of Philosophy Quarterly.* 27 (4): 359-378.

Pelto P. J. and G. H. Pelto. 1997. Studying Knowledge, Culture and Behavior in Applied Medical Anthropology. *Medical Anthropology Quarterly.* 11 (2): 147-63.

Petrie, K. J., L.A. Jago and D. A. Devcich. 2007. The Role of Illness Perceptions in Patients with Medical Conditions. Lippincott Williams and Wilkins: *Current Opinion in Psychiatry.* 20 (2-3): 163-167.

Qadeer, Imrana and Dunu, Roy. 1989. Work, Wealth and Health: Sociology of Workers' Health in India. *Social Scientist.* 17 (5/6): 45-92.

Qadeer, Imrana. 1995. Primary Health Care: A Paradise Lost. *IASSI Quarterly.* 14 (1&2): 1-20.

Qadeer, Imrana. 2002. Draft National Health Policy 2001-I: Debt Payment and Devaluing Elements of Public Health. *Economic and Political Weekly.* 37 (1): 12-16.

Radley, Alan., Julianne, Cheek and Christian Ritter. 1997. The Making of Health: A Reflection on the First 10 years in the Life of a Journal. In *Health: An Interdisciplinary Journal for the Social Study of Health, Illness and Medicine,* p.141.London: Thousand Oaks.

Ramaswamy, Geeta. 2005. *India Stinking: Manual Scavenging in Andhra Pradesh and their Work.* Chennai: Navayana.

Reader, George. andMary E. W. Goss. 1959. The Sociology of Medicine. In Robert K. Merton., Leonard and Leonard S. Cotterll *Sociology Today: Problems and Prospects.* pp. 229-246. New York: Basic Books.

Reeler, A. V. 2000. Anthropological Perspectives on Injections: A Review. *Bulletin World Health Organization.* 78 (1): 135-43.

Richard, L. Gregory. 2001. Perceptions of Knowledge. *Nature.* 21 (1 March): 410.

Rosenberg, Charles E. 1989. Disease in History: Frames and Framers. *Milbank Quarterly.* 67 (Supplement 1): 4-5.

Ruiz-Beltran M, and J.K. Kamau. 2001. The Socio-economic and Cultural Impediments to Well-being Along the US-Mexico Border. *Journal of Community Health.* 26 (2): 123-32.

Russel, R. V. and Rai Bahadur Hira Lal. 1916. *The Tribes and Castes of the Central Provinces of India.* Vol. 4. London: Mcmillan.

Sabharwal, N. S. 2004. Caste, social exclusion and health deprivation in India. Indian Institute of Dalit Studies.

Sachchidananda. 2002. *People at the Bottom: A Portrait of Scavengers.* New Delhi: Concept Publications.

Sahu, S. K. 1986. Social Dimension of Health of Tribals in India: A Case Study of Oraons of Orissa. In Buddhadeb Chaudhuri (ed.) *Tribal Health: Socio-cultural Dimensions,* New Delhi: Inter-India Publications.

Sahu, S.K. 1980. *Health Culture in Transition: A Case Study of Oraon Tribe in Rural and Industrial Nexus.* New Delhi: Khama Publishers.

Seale, C., and Charteris-Black, J. 2008. The interaction of age and gender in illness narratives. *Social Science & Medicine,* 62(10), 2577–2590

Sengupta, Nirmal. 1980. *Destitutes and Development: A Study of the Bauri Community in the Bokaro Region.* New Delhi: Concept Publishing Company.
Shah, A M. 1998. *The Family in India: Critical Essays.* New Delhi: Orient Longman.
Shah, A. M. 1996. Is the Joint Family Disintegrating. *Economic and Political Weekly.* 31 (9): 537-542.
Shah, A.M. 2002. The Dalit 'Category' and Its Differentiation. *Economic and Political Weekly.* 37 (14): 1317-1318.
Shah, A.M. and Aneeta Ahluwalia. 1970. Role of Sociology in Medical Education and Research. *Economic and Political Weekly.* 5 (17): 705-10.
Shahrawat, R. and K. D. Rao. 2011. Insured Yet Vulnerable: Out-of-Pocket Payments and India's Poor. *Health Policy Plan.* 27 (3): 213-21.
Shyamlal. 1981. The Occupational Mobility among Urban Bhangis. *Eastern Anthropolgist.* 11 (1): 11-20.
Shyamlal. 1984. *The Bhangis in Transition.* New Delhi: Inter India Publications.
Shyamlal. 1992. *The Bhangi: A Sweeper Caste Its' Socio-economic Portraits with Special Reference to Jodhpur City.* Bombay: Popular Prakashan.
Sigerist, Henry E. 1952. *History of Medicine.* New York: Oxford University Press.
Singh, Kumar Singh. 1995. *People of India: Daman and Diu.* Anthropological Survey of India. pp. 74-75. New Delhi: Popular Prakashan Private Limited, XIX.
Slappa, I.P.D. 1976. *Committee on Improvement of Living and Working Conditions of Sweepers and Scavengers.* Bangalore: Government of Karnataka.
Smith, Judith A. R. N. 1981. The Idea of Health: A Philosophical Inquiry. *Advances in Nursing Science.* 3 (April): 43-50.
Somayeh, Mohammadi and Banirostam, Touraj. 2015. A Perceptual Meta-model based on the Ontology of Mental Models. *The International Journal of Humanities and Social Studies.* 3 (11): 112.
Sonia, Marrone. 2007. Understanding Barriers to Health Care: A Review of Disparities in Health Care Services among Indigenous Populations. *International Journal of Circumpolar Health.* 66 (3): 188.
Srinivas, M. N. 1942. *Marriage and Family in Mysore.* Bombay: New Book Company.
Srinivas, M. N. 1952. Religion and Society among the Coorgs of South India. Oxford: Clanderon Press.
Srinivas, M. N. 2003. An Obituary on Caste as a System. *Economic and Political Weekly.* 38 (5): 455-459.
Srinivas, M. N., A.M. Shah and E. A. Ramaswamy. 2002. *The Fieldworker and the Field.* New Delhi: Oxford University Press.
Srinivasan, K. and S.K. Mohanty. 2008. National Family Health Survey 3- Household Deprivation and Its Linkages with Reproductive Health Utilization. *Economic and Political Weekly.* 43 (48): 67-76.
Srivastav, B N. 1997. *Manual Scavenging in India: A Disgrace to the Country.* New Delhi: Concept Publishing Company.Srivastava, Vinay Kumar. (ed.) 2004. *Methodology and Fieldwork.* New Delhi: Oxford University Press.
Suchman, Edward A. 1965. Stages of Illness and Medical Care. *Journal of Health and Human Behaviour.* 6 (3): 114-28.
Sujatha, V. 2003. *Health by the People: Sociology of Medical Lore.* New Delhi: Rawat Publications.
Sunder, Ramanmani. 1995. *Household Survey of Health Care Utilisation and Expenditure.* The Working Paper No. 53. New Delhi: National Council for Applied Economic Research.
Thorat, S., and Sabharwal, N. S. 2006. Caste, social exclusion and health [Working papers/essays on social exclusion; exact title varies across editions—commonly cited IIDS/UNICEF work]
Thorat, S., and Newman, K. S. 2010. *Blocked by caste: Economic discrimination in modern India.* Oxford University Press.
van Doorslaer, E., O'Donnell, O., *et al.* 2007. Catastrophic payments for health care in Asia. *Health Economics, 16*(11), 1159–1184. https://doi.org/10.1002/hec.1209

Venkatesan, V. 2005. A Case for Human Dignity. *Frontline*. 22 (12). June-4-17.

Vijayanath, V. *et al.* 2010. Caste and Health in Indian Scenario. *Journal of Punjab Academy of Forensic Medicine and Toxicology*.(Online). 10 (1): 42-43. Acccesed on 18 May 2016, www.journals.pafmat.com/20101_11.php.

Wale, N. J. (1976). Sociology and health care. Macmillan.

Wan, T.T.H. and S.J. Soifer. 1974. Determinants of Physician Utilisation: A Causal Analysis. *Journal of Health and Social Behaviour.* 15 (2): 100-108.

Webster, A. Merriam. 1993. *Webster's Third New International Dictionary*. Chicago: Encyclopedia of Britannica.

Weiss, G. L., and Lonnquist, L. E. 1997. The sociology of health, healing, and illness (3rd ed.). Prentice Hall.

World Health Organization (WHO). 1978. *Declaration of Alma-Ata*. WHO/UNICEF International Conference on Primary Health Care, Alma-Ata, USSR, 6–12 September 1978.

World Health Organization. 2009. *Self-care in the Context of Primary Health Care: Report of the Regional Consultation Bangkok.* Thailand 7–9, January 1-2.

Yamin, Alicia Ely. (ed.) 2005. *Learning to Dance: Advancing Women's Reproductive Health and Well-Being from the Perspectives of Public Health and Human Rights*. p.1, cited from Chandirma Chatterjee and Gunjan Sheoran *Vulnerable* Groups In India Mumbai: CEHAT.

Ziyauddin, K.M. 2022. Situating Hadis' Occupation and Caste. CASTE: A Global Journal on Social Exclusion 3 (2), 263-284

Zola, Irving K. 1973. Pathways to the Doctor - from Person to Patient. *Social Science and Medicine*. 7 (9): 677-689.

Zurbrigg, Sheila. 1984. *Rakku's Story: Structures of Ill Health and the Source of Change*. Bangalore: Centre for Social Action.

The manufacturer's authorised representative in the EU is Springer Nature Customer Service Centre GmbH, Europaplatz 3, 69115 Heidelberg, Germany. If you have any concerns regarding our products, please contact ProductSafety@springernature.com

Printed and bound by CPI Group (UK) Ltd, Croydon, CR0 4YY
30/09/2026
02225303-0005